T0245244

CAMBRIDGE LIBRARY COLLECTION

Books of enduring scholarly value

Life Sciences

Until the nineteenth century, the various subjects now known as the life sciences were regarded either as arcane studies which had little impact on ordinary daily life, or as a genteel hobby for the leisured classes. The increasing academic rigour and systematisation brought to the study of botany, zoology and other disciplines, and their adoption in university curricula, are reflected in the books reissued in this series.

A Sketch of the Life and Labours of Sir William Jackson Hooke

Sir William Jackson Hooker (1785–1865) was an eminent British botanist who is best known for expanding and developing the Royal Botanic Gardens at Kew into a leading centre of botanic research and conservation. After undertaking botanical expeditions to Iceland and across Europe, he was appointed Regius Professor of Botany at Glasgow University in 1820, where he proved to be a popular lecturer and established the Royal Botanical Institution of Glasgow. In 1841 Hooker was appointed the first Director of the Royal Gardens at Kew, a position he held until his death. This volume, written by his son, the equally renowned botanist Sir Joseph Hooker (1817–1911) and first published in 1903, provides an intimate biography of his life. Hooker's botanic expeditions, his experiences at Glasgow, and relations between leading members of the scientific community are recounted, together with vivid descriptions of his labours and improvements at Kew.

Cambridge University Press has long been a pioneer in the reissuing of out-of-print titles from its own backlist, producing digital reprints of books that are still sought after by scholars and students but could not be reprinted economically using traditional technology. The Cambridge Library Collection extends this activity to a wider range of books which are still of importance to researchers and professionals, either for the source material they contain, or as landmarks in the history of their academic discipline.

Drawing from the world-renowned collections in the Cambridge University Library, and guided by the advice of experts in each subject area, Cambridge University Press is using state-of-the-art scanning machines in its own Printing House to capture the content of each book selected for inclusion. The files are processed to give a consistently clear, crisp image, and the books finished to the high quality standard for which the Press is recognised around the world. The latest print-on-demand technology ensures that the books will remain available indefinitely, and that orders for single or multiple copies can quickly be supplied.

The Cambridge Library Collection will bring back to life books of enduring scholarly value (including out-of-copyright works originally issued by other publishers) across a wide range of disciplines in the humanities and social sciences and in science and technology.

CAMBRIDGE UNIVERSITY PRESS

Cambridge, New York, Melbourne, Madrid, Cape Town, Singapore,
São Paolo, Delhi, Dubai, Tokyo

Published in the United States of America by Cambridge University Press, New York

www.cambridge.org
Information on this title: www.cambridge.org/9781108019323

This edition first published 1903
This digitally printed version 2010

ISBN 978-1-108-01932-3 Paperback

A Sketch of the Life and Labours of Sir William Jackson Hooke

Late Director of the Royal Gardens of Kew

Joseph Dalton Hooker

A SKETCH OF THE LIFE AND LABOURS

OF

SIR WILLIAM JACKSON HOOKER, K.H.

D.C.L. Oxon., F.R.S., F.L.S., Etc.

HENRY FROWDE, M.A.
PUBLISHER TO THE UNIVERSITY OF OXFORD
LONDON, EDINBURGH
NEW YORK

Walter L. Colls, Ph. Sc.

W. J. Hooker

A Sketch of the Life and Labours

of

Sir William Jackson Hooker, K.H.

D.C.L. Oxon., F.R.S., F.L.S., &c.

LATE DIRECTOR OF THE ROYAL GARDENS OF KEW

BY HIS SON

JOSEPH DALTON HOOKER

OXFORD

AT THE CLARENDON PRESS

1903

ADDITIONAL NOTICES, EMENDATIONS, AND CORRECTIONS

p. ix, on footnote. Wallich in founding the genus *Hitchenia*, after Mr. Hitchin, erred in the spelling.

p. xiii, l. 7—*for* Scheuzeria *read* Scheuchzeria.

p. xxv, l. 12 from bottom—*omit* the perplexing issue will be described later on.

p. xxvi, l. 3—*for* Thibet *read* Tibet, also in l. 2 from bottom of preceding page.

p. xxxi, l. 18—*for* Jordan Hill *read* Jordanhill.

p. xl, l. 7—the list here alluded to is given in the Appendix (A) following this Sketch in the 'Annals of Botany,' Vol. XVI.

p. lv, footnote [1]—*replace by* the Grounds of West Park are now occupied by the Sewage Works of Kew and Richmond.

p. lxiii, l. 15—*for* Wilfred *read* Wilford; his mission to Japan preceded that of Oldham as did that of Barter in W. Africa precede that of Mann. Miller was associated with Kirk in Livingstone's expedition of 1863, and Macgillivray and Milne served together in H.M.S. Herald's voyage.

p. lxviii. l. 3 from bottom—Professor Oliver in 1864 succeeded Mr. A. Black, who was the first Curator of the Herbarium, having been appointed in 1853. Mr. Black's health failing he retired in 1864, became Superintendent of the Botanical Garden of Bangalore and died in 1865.

p. lxxix, footnote [2]—*omit* 'and Baker' in the last line.

p. lxxxi, l. 17—*for* Thibet *read* Tibet.

p. lxxxii, in footnote—*for* Cowfield *read* Cowfold.

p. lxxxiii—I regret the omission of the name of Miss Bromfield amongst the contributors to the Herbarium of Kew. In 1853 that lady presented to Kew the botanical library collections and MSS. of her deceased brother, Dr. W. Arnold Bromfield, F.L.S., of Ryde, Isle of Wight. This gift consisted chiefly of British, N. American, West Indian, and Palestine plants, together with about 300 botanical works, including fine copies of the best editions of some of the early botanical writers and costly folios of later authors, many of them of great rarity.

p. lxxxv, l. 17—after Church—*enter*—modelled by his nephew, Sir Reginald Palgrave, K.C.B., late Clerk of the House of Commons.

A SKETCH OF THE LIFE AND LABOURS

OF

SIR WILLIAM JACKSON HOOKER.

(*With Portrait*).

CHAPTER I.

NORWICH AND HALESWORTH, 1785–1820.

WILLIAM JACKSON HOOKER was born in St. Saviour's parish, Norwich, on July 6, 1785. He was the younger of two sons, the only children of Joseph and Lydia Hooker, of that city. His father was a native of Exeter, the home of many generations of the Devonshire Hookers [1], where he had been a confidential clerk in the house of Baring Brothers, wool-staplers, with whose family his was distantly connected. From Exeter he went to Norwich, and into business there, where he had a collection of 'Succulents,' the cultivation of which class of plants was a favourite pursuit of many of his fellow citizens [2]. He was mainly a self-educated man, and a fair German scholar. My father's mother was a daughter

[1] Descendants of John Hooker, alias Vowell, First Chamberlain of Exeter and member for the city, editor of Holinshed's Chronicles, for which he wrote the history of the Irish Parliament and translated the Irish Histories of Giraldus Cambrensis, &c. He was uncle of Richard Hooker, whom he sent to college. My grandfather was seventh in descent from John, whose ancestors (fide Heralds' College) date back for six generations to a Seraph Voell, of Pembroke; but except John, Richard, and a John who was M.P. for Exeter, temp. Edward V, Richard III and Henry VII, not one of the long line, in so far as I know, emerged from obscurity.

[2] The best known of these collections was that of Thomas Hitchin, a dyer of Norwich, after whom Wallich named the noble Burmese plant *Hitchinia glauca*. In 1882 I could hear of but one collection remaining in the city, that of Dr. Masters, since dispersed, some of the contents coming to Kew.

of James Vincent, Esq., of Norwich, a worsted manufacturer, grandfather of George Vincent[1], one of the best of the Norwich School of artists, and whose works are now much sought for. Thus my father presumably derived his love of plants from his father's side, and his artistic powers from his mother's.

Of my father's early childhood I know no more than that he went to the Norwich Grammar School, under the then well-known pedagogue, Dr. Foster, but the records of that school having been destroyed it is impossible to say what progress he made there; at home he devoted himself to entomology, drawing, and reading books of travel and natural history. When only four years old he inherited the reversion to a fair competency in landed and personal property in Kent, through the death of his cousin and godfather, William Jackson, Esq.[2], of Canterbury, a young man of great promise. After leaving school he was sent to reside with a Mr. Paul, of Starston (a village on the borders of Suffolk), a gentleman farmer, who instructed sons of the landed gentry in the management of estates. Early in life he devoted himself to ornithology, visiting the Broads and sea-coast of Norfolk, which abounded in rare birds, shooting, stuffing, and drawing them, besides learning their habits and songs. Sixty years later he knew the birds in Kew Gardens by the eye and the ear, and in a manner which surprised me. Though a keen ornithologist and as keen an entomologist, he was almost morbidly averse from taking life; he never shot for sport or for the pot, and many years afterwards when instructing me in entomology he was ever urging me to kill with the least suffering, and never to take more specimens than were necessary. His was one of those temperaments that later in life

[1] George Vincent was well educated and brought up, but lost himself. My father, his cousin, vainly endeavoured to trace his end in London.

[2] He was killed in 1789, being thrown from his horse at his father's door; see Hasted's Kent, iv. 427, and, for a long *éloge*, the Gentleman's Magazine, lxii (1790), 859. A sermon is to this day annually preached, in memory of him, in St. Mildred's Church, Canterbury, where is also his monument by the sculptor Bacon.

could not look on blood without a feeling of faintness, or on a wax model of the human face with equanimity.

That his entomological pursuits were, when still in his teens, appreciated by the veteran Kirby is evidenced by the latter having in 1805 dedicated to him and his brother a species of *Apion* with these words: 'I am indebted to an excellent naturalist, Mr. W. J. Hooker, of Norwich, who first discovered it, for this species. Many other nondescripts have been taken by him and his brother, Mr. J. Hooker, and I name this insect after them, as a memorial of my sense of their ability and exertions in the service of my favourite department of natural history [1].'

I do not know the age at which my father took up botany. The first evidence of his having done so is the fact, that he was the discoverer in Britain in 1805 of a very curious moss, *Buxbaumia aphylla*; but it may be inferred from this and from his correspondence with Mr. Turner (which I possess) that he had at the age of twenty-one thoroughly studied not only the flowering plants, but the mosses, Hepaticae, lichens [2], and fresh-water Algae of Norfolk. The *Buxbaumia* he took to his friend Dr. (afterwards Sir James) Smith [3], of Norwich, the possessor of the Linnean herbarium, who advised him to send specimens to Mr. Dawson Turner, F.R.S. [4], of Great

[1] Transactions of the Linnean Society, vol. ix (1808), p. 70.

[2] In a letter dated March, 1806, he mentions having a cabinet made for his collection of lichens with twenty-eight or thirty drawers, each two inches deep with thirty-six partitions, in which to place cards with mounted specimens.

[3] In 1808 Sir James Smith dedicated a genus of mosses to him in the following words: 'I have great pleasure in dedicating this genus (*Hookeria*) to my young friend, William Jackson Hooker, F.L.S., a most assiduous and intelligent botanist, already well known by his interesting discovery of *Buxbaumia aphylla*, as well as by his scientific drawings of *Fuci* for Mr. Turner's work; and likely to be far more distinguished by his illustrations of the difficult genus *Jungermannia*, to which he has given particular attention' (Trans. Linn. Soc. ix. 1808, 275). The plate accompanying Sir James Smith's paper is of four species of the genus, signed 'W. J. Hook. delin^{t}.'

[4] Mr. Turner was a partner in Gurney's Bank, Great Yarmouth, of which his father was one of the founders. He was eminent as a scholar, botanist, antiquarian, and bibliophile. His collection of royal autographs and his illustrated copy of Blomfield's Norfolk are in the British Museum.

Yarmouth, author of 'Muscologiae Hibernicae Spicilegium,' and, with L. W. Dillwyn, F.L.S., of 'The Botanist's Guide through England and Wales.' This he did, and it was immediately followed by an invitation from Mr. Turner to visit him, which led to the colouring of his future life.

In 1806, when only four months over his majority, my father was elected a Fellow of the Linnean Society, probably the youngest individual so honoured. In the same year he visited London, and was introduced to Sir Joseph Banks, Konig, Brown, and other naturalists. The years 1806–9 were passed between Norwich, Yarmouth, and London, with intervals of travelling in Scotland and Iceland. In London he had rooms in Frith Street, Soho, to be near the British Museum, Linnean Society's rooms, his friends, R. Brown, Leach, Konig, Edward Foster, Macleay, and above all the Banksian library and collections, and Sir Joseph Banks himself, who treated him with great kindness, stimulating his zeal as a naturalist and his desire to travel. At Yarmouth, where he was a frequent guest for protracted periods, he devoted himself mainly to aiding Mr. Turner in his great work, the 'Historia Fucorum[1],' of which aid the latter makes frequent grateful mention in his correspondence with Mr. Borrer. During the same period he was occupied with preparing his 'British Jungermanniae' for publication[2], and in studying Buchanan-Hamilton's Nepal mosses in Sir James Smith's herbarium, upon some of which he wrote his first published paper. It is entitled 'Musci Nepalenses,' and was read before the Linnean Society in June, 1807 (Linn. Trans., ix. 1807, pp. 26-8, with three plates).

In 1807, when botanizing in the neighbourhood of Yarmouth, he was bitten by a viper. Fancying he had been pricked by a thorn he paid no heed to the pain till giddiness came on,

[1] Of the 258 coloured plates of this work, 231 are inscribed 'W. J. H., Esqr. delt.' in minute letters; 12 signed 'M. T.,' or 'D^{na} T., are by Mrs. Turner; 7 by Miss Hutchins, of Bantry; 2 by Professor Martens, of Bremen, and 1 by Sir Thomas Frankland.

[2] Writing to Mr. Turner in 1808 he mentions that Dr. Smith had lent him the whole Linnean collection of *Jungermanniae* for study, together with his own.

under which he succumbed. After lying for some time in a state of collapse[1] he was accidentally found by some friends, who carried him to Mr. Turner's, where violent fever supervened, followed by a tedious illness. On recovery he started with Mr. and Mrs. Turner on a botanical tour in Scotland. Their route was, first, Croft in Yorkshire, visiting the Rev. James Dalton, F.L.S., the discoverer of the *Scheuzeria* in England, after whom the moss *Daltonia* is named; then Carlisle, Branksome, Melrose, Edinburgh, the Falls of Clyde, Glasgow, Dumbarton, Luss, Ben Lomond—ascended in cloud and rain, guided by the Rev. Dr. Stuart, of Luss, an excellent botanist, a friend of Lightfoot, and the translator of the New Testament in Gaelic. Thence they proceeded to Inverary, Loch Awe, Oban, Mull, Ulva, Staffa, Fort William, ascending Ben Nevis in terrible weather, Fort Augustus, Elgin, visiting Mr. Brodie of Brodie, F.R.S., the discoverer of *Moneses* and other rare plants in Scotland. Thence to Loch Tay, ascending Ben Lawers twice, Killin, ascending Ben Cruachan, Craighalliach and Ben More, Stirling, Edinburgh, and Newcastle, visiting Mr. J. Winch, F.L.S., author of the 'Geographical Distribution of Northumbrian Plants,' and Mr. J. Thornhill, of Gateshead, a good local botanist; thence to Darlington on a visit to Mr, Backhouse, banker, who showed them *Cypripedium Calceolus*, and so back to Yarmouth.

In 1808 my father undertook a much longer journey in Scotland, accompanied by his friend Mr. Borrer[2]. On this occasion he reascended Ben Lawers, Ben Lomond, Ben Cruachan, and Ben Nevis, and for the first time Shichallion, Ben Hope, and Ben Loyal. After visiting Mr. Brodie of Brodie, they went to Caithness and the Orkneys, returning to Sutherland. In a letter to Mr. Turner he thus describes their reception in Sutherland: 'We did not leave North Sutherland with the good wishes of the inhabitants, at least

[1] The symptoms, as described in a letter from Turner to Borrer, were dreadful giddiness, pain about the navel, shivering, drowsiness, vomiting, purging, and exhaustion.

[2] William Borrer, Esq., of Henfield, Sussex, F.R.S., F.L.S., died 1862, aged 81; the Nestor of British botanists.

the lower classes of them, most of whom took us for French spies, or, what is worse in their estimation, sheep-farmers. Daniel Forbes, who so often acted as our guide, was advised by some to conduct us by the worst way possible; by others he was told that he might be better employed. Our lad heard some saying that we ought to be flogged and sent out of the country. They have not the least idea of persons travelling for mere curiosity, and could not be persuaded that we were not come to do them some ill.' Crossing Sutherland and Cromarty, they went by Moida and Lairg to Skye, where they found the *Eriocaulon*, and to the remarkable and little-visited cave of Slock Altramins. Recrossing the Sound to Glenelg, they proceeded to visit Sir George McKenzie at Coul, and Lord Seaforth at Brahan Castle, and again Mr. Brodie of Brodie, returning by Aviemore, Killiecrankie, and Edinburgh[1] to Norwich.

The journey through the North of Scotland was performed mainly on horses or ponies, and the difficulties met with were such as can now be experienced only in the out-of-the-way parts of the globe. My father made copious pencil sketches and kept a journal, which he was vainly urged by his friends to publish. I have no idea what became of it. The only recorded botanical result of the journey was the discovery of a new *Andreaea* (*A. nivalis*, Hook.) on the summit of Ben Nevis; which probably prompted the writing of his second published paper, 'Some Observations on the Genus *Andreaea*,' read before the Linnean Society in May, 1810 (Linn. Trans., x. 381, tab. xxxi).

In 1809 Sir Joseph Banks, hearing of an opportunity for a naturalist visiting Iceland, where he himself had been in 1772, suggested my father's taking advantage of it. This he did, and all the more eagerly from having as a boy read 'Van Troil's Letters on Iceland,' with a longing to visit the hot springs and volcanoes therein described. The opportunity was the dispatch of a vessel, the *Margaret and Anne*, with

[1] It was probably on this occasion that my father became one of the founders of the Wernerian Society of Edinburgh, the memoirs of which, commenced in 1808, were concluded in 1832, in six volumes.

a letter of marque, chartered by a London firm, Messrs. Phelps & Co., for the purpose of obtaining a cargo of tallow. The venture was a risky one, for Denmark, to which country Iceland belonged, was at war with England, and the firm were enticed to undertake it by a Danish prisoner of war, Jorgen Jorgensen by name, who was now for the second time about to break his parole and accompany the ship in the interests of the firm. The *Margaret and Anne* sailed June 2, and on arriving June 21 at Reikevik Jorgensen, finding that commerce with England was prohibited, effected a revolution in the island, proclaimed its independence of the Danish crown and himself its 'Protector,' imprisoned the governor, Count Tramp, erected a fort armed with six guns, equipped troops, remodelled the laws, established representative government and trial by jury, reduced the taxes, and raised the salaries of the clergy; all without shedding a drop of blood, or an attempt at resistance on the part of the people[1]!

On his arrival at Reikevik my father received a hearty welcome from the Stiftsamptman (Icelandic governor of the island), to whom he had brought from Sir Joseph Banks a letter of introduction, together with a handsome present of books, engravings, &c. The delight of the old man on receiving these was affecting; he spoke of Sir Joseph with veneration, describing his philanthropic efforts to avert the

[1] An account of the career of this extraordinary man is given in his Autobiography, published anonymously in Ross's Hobart Town Almanack for 1835; and is retold in a little work entitled 'The Convict King, being the Life and Adventures of Jorgen Jorgensen, Monarch of Iceland, Naval Captain, Revolutionist, British Diplomatic Agent, Author, Dramatist, Preacher, Political Prisoner, Gambler, Hospital Dispenser, Continental Traveller, Explorer, Editor, Expatriated Exile, and Colonial Constable, retold by James Francis Hogan': 12mo, London, Ward and Downey, 1891. What most concerns botany in Jorgensen's career are the facts that he served as a seaman under Capt. Flinders, R.N., in his voyage to Terra Australis (1802-5), with Robert Brown as botanist, and J. Franklin (afterwards Sir John, the Arctic traveller) as midshipman; and that it was through the exertions of Mrs. Fry, Sir Joseph Banks, and my father, that the sentence of death passed on Jorgensen in 1825 was commuted into penal servitude for life in Tasmania, where I saw him in 1840. He died there in that or the following year, his fellow voyager, Sir John Franklin, being governor of the colony at the time! See 'Tour in Iceland,' by W. J. H., for details of Jorgensen's acts, &c.

famine that threatened the Icelanders at the beginning of the war, when the activity of our cruisers intercepted their supplies of food from Denmark and Norway [1], adding that Sir Joseph had obtained the release of Danish prisoners in England, and at his own expense furnished them with the means of living and returning to their homes.

As may be taken for granted, under such circumstances every facility was given to the visitor for travelling to the most interesting places in the island, Thingewalla, the Geysers, Skalholt, Reykholt, &c., and for making collections and observations on natural history.

On August 6, H.M.S. *Talbot* anchored in Reikevik harbour, when her commander, the Hon. Captain Jones, promptly deposing and making a prisoner of Jorgensen, replaced Count Tramp in the governorship.

On August 25, after bidding adieu to his kind friend the Stiftsamptman, who gave him a valuable collection of Icelandic books, my father embarked on his return voyage in the *Margaret and Anne*. On this occasion the vessel carried besides the passengers and crew some Danish prisoners of war, and she was ordered by Captain Jones to sail in company with the *Orion* [2], now a prize of the *Talbot*, carrying Mr. Jorgensen and another party of Danish prisoners. The two ships left in the afternoon, but the *Orion* becoming suddenly becalmed could not proceed till the following day. The *Margaret and Anne*, on the other hand, being favoured by the wind, pursued her voyage till the morning of the 27th, when being twenty leagues from the land, in a dead calm, she was discovered to be on fire. Being loaded with oil and tallow, the progress of the flames was rapid; smoke burst out at once from all the hatches, and to add to the horror of the situation, she did not

[1] Sir Joseph Banks being himself a Privy Councillor obtained an Order in Council, dated Feb. 10, 1810, strictly forbidding acts of hostility against the poor and defenceless colonies of the Danish dominion, and permitting them to trade with the parent-country, unmolested by British cruisers.

[2] The *Orion* was a Danish ship of war, that had brought Count Tramp to Iceland a few weeks before the arrival of the *Margaret and Anne*, which, in virtue of her letter of marque, had, under Jorgensen's orders, seized her as a prize.

carry boats enough to hold the number of souls on board. All attempts to subdue the fire were vain, when providentially a rescuer appeared on the horizon. This was the *Orion* with the irrepressible Jorgensen[1] on board, who, to enable that vessel to rejoin her consort, had insisted on being allowed to run her through a dangerous passage between the reefs and the mainland of Reikevik harbour, and who by thus saving a day saved the lives of all hands on the burning ship, whom he carried back to Reikevik.

My father's description[2] of the progress of the conflagration, as seen from the *Orion*, is graphic—of the flames seizing the sails and rigging, of the falling of the masts, of the discharge of the guns, and of the reduction of a ship of 500 tons burthen, worth £25,000, to a hull with cataracts of blazing oil and tallow pouring over its sides.

Unfortunately the fire broke out in a part of the ship where his collections were stowed, and he lost everything but a few weeks of his journal, the clothes he stood in, and an Icelandic lady's wedding dress[3], which the ship's steward flung into the boat as she shoved off from the burning wreck.

The fire was proved to have been planned before leaving Reikevik by some of the Danish prisoners, two of whom had lit it in the previous night. A search in the bedding of the prisoners in the *Orion* resulted in finding combustible materials, no doubt secreted for the same object.

On her return to Reikevik Captain Jones offered my father a passage home on board the *Talbot*, which he gladly accepted. The voyage was a tempestuous one of sixteen days' duration, during which the *Talbot* lost her foremast. She arrived in Leith roads on September 20.

[1] Jorgensen had proved himself to be a first-rate seaman, with all the qualities of a commander, when serving under Captain Flinders; and subsequently in 1807, as a captain in the Danish navy, when in a ship with eighty-three hands and twenty-eight guns, he engaged for three-quarters of an hour the British sloop *Sappho*, with 120 men. On this occasion he was taken prisoner and put on parole, which he twice broke as stated above, in making this and a former visit to Iceland in the interests of Messrs. Phelps & Co.

[2] See Tour in Iceland, vol. i, pp. 362–4.

[3] Now in the Victoria and Albert Museum, South Kensington.

Soon after his return, and yielding to the wishes of his friends, he commenced writing his 'Journal of a Tour in Iceland.' On hearing of this Sir Joseph Banks most liberally offered him the use of his own manuscript journal, and various other papers relating to the island, together with the magnificent drawings of the scenery, dresses of the inhabitants, &c., which were made by the artist who accompanied him in his voyage thither in 1772. With these materials, his own journal of four weeks out of the twelve which he passed in the island, and a retentive memory, refreshed by a reference to all available works and all documents relating to the revolution, he compiled and printed, for *private distribution only*, in 1811, an 8vo volume of upwards of 400 pages and four plates. Sir Joseph Banks was so pleased with it that he induced my father to reproduce it for publication. The second edition with additions, in two volumes, with two maps and four plates, dedicated to Sir Joseph, appeared in 1813, and is to this day a standard work. A *résumé* of its contents may be welcome to those interested in the author's career. Volume i contains the history and present condition of Iceland, its productions, institutions, commerce, &c., followed by his 'Recollections of Iceland' in journal form. Volume ii consists of six appendices: (1) details of the Icelandic Revolution, drawn up with singular impartiality; (2) proclamations relating to it; (3) Hecla and the volcanic mountains of Iceland; (4) Odes and Letters presented by the literati of Iceland to the Right Honourable Sir Joseph Banks and the Honourable Captain Jones; (5) a list of Icelandic plants; (6) Danish ordinances concerning the trade of Iceland.

Reverting to the destruction of his collections, my impression is that the loss to science of the cryptogamic plants was the most serious, for he was a keen student of mosses, Hepaticae, lichens, and both marine and fresh-water Algae[1], and had gained invaluable knowledge on them during his excursions

[1] When only twenty-one years old he was in correspondence, on the subject of fresh-water Algae, with Mr. Dillwyn for whose British Confervae he made drawings of species discovered by himself in Norfolk.

in the east and north of England, and especially during his two extended Scottish tours. Of flowering plants he probably added but few to the list he gives of 359 species taken from Zoega's 'Flora Islandica,' published in 1772, with the addition of twenty-two from his own observations and Sir George McKenzie's collections. Babington, in his very valuable 'Review of the Flora of Iceland' (Journ. Linn. Soc., xi. 1870, 282), enumerates 433 Icelandic flowering plants, which is an increase of 96 species.

The years immediately following my father's return from Iceland (1809–12) were the most embarrassing of his life. His unquenchable longing to travel in the tropics was kept alive by Banks's earnest endeavours to find him a fitting opportunity. On the other hand his botanical friends were unanimous in urging him to remain at home, publish his Icelandic and Scottish journals, continue his aid to Mr. Turner on the 'Historia Fucorum,' and, above all, proceed with his 'British Jungermanniae,' his drawings and analyses of which were of unrivalled beauty, and his contemplated 'Muscologia Britannica.' Meanwhile Mr. Turner, with real desire to benefit his young friend, induced him in 1809 to join in partnership with himself and Mr. Paget of Yarmouth (father of the late Sir James Paget) in a brewery at Halesworth, reside there, and undertake the management of a business for which he had neither experience nor inclination. This did not check either his botanical ardour or his desire to visit the tropics. In 1810 he sold his landed property and determined to accept an invitation, which Sir Joseph had procured for him, of accompanying Sir Robert Brownrigg, G.C.B., the newly appointed Governor of Ceylon, to that island. To this end he appointed his father locum tenens at the brewery and proceeded to London, where amongst other preparations for the undertaking he made, at the Museum of the India House, reduced pen and ink sketches from upwards of 2,000 folio drawings of Indian plants[1], which had been executed by native

[1] These drawings, now in the Herbarium of the Royal Gardens, Kew, are

artists in the Botanical Gardens of Calcutta under Dr. Roxburgh's directions.

To his bitter disappointment this opportunity had to be put aside, for disturbances followed by a rebellion had broken out in Ceylon that would have rendered travelling in the island impossible. One more chance presented itself in 1813. Through his intercourse with Dr. Horsfield[1], the Keeper of the India House Museum, his attention was turned to Java, where that officer had resided under Sir Stamford Raffles's rule, and had made magnificent collections. Sir Joseph Banks encouraged the idea of his going there, and prevailed on Lord Bathurst, the President of the Board of Trade, to remunerate him if he would send living plants to Kew, and procure information regarding the cultivation of spice-bearing trees in the Dutch East Indies. But disappointment still pursued him. The climate of Java was reported to be notoriously malarious, and Banks's own experience of it, as narrated in Cook's 'First Voyage[2],' was cited in evidence. For there, not only had Banks been extremely ill, but Dr. Solander had been at death's door; and Mr. Parkinson his artist, the two Otaheitans in his suite, Mr. Green the astronomer, Mr. Monkhouse the surgeon, and Mr. Spring had all died from the effect of the climate at or shortly after leaving Batavia. No wonder that the entreaties of his parents and friends prevailed, notwithstanding Banks's well-founded assurance that the climate of Java itself was as exceptionally good as that of Batavia was bad. My father was hence compelled to confine his wanderings to nearer home, adding gardening to his pursuits, and this with some success, for he was the first to flower *Cattleya labiata* in his little stove in 1818, and he also flowered *Musa coccinea* and other tropical plants.

duplicates (exact copies) of the originals in the Royal Botanic Gardens, Calcutta. The reductions by my father occupy ten duodecimo volumes, also in the Kew Library.

[1] Thomas Horsfield, M.D., F.R.S., F.L.S., Keeper of the India House Museum, 1820–59. For a sketch of his travels see Brown and Bennett, Plantae Javanicae Rariores, postscript, pp. i–xvi.

[2] The *Endeavour* lost, from malarial fever or its effects, seven persons in Batavia, and twenty-five after leaving that port.

In 1813, owing to the illness of his only brother[1], my father spent five months with him in Devonshire and Cornwall, which counties he diligently explored for Musci, Hepaticae, and lichens especially. The Trinity House yacht having been placed at his disposal, he visited the Scilly Islands, whence he writes to Mr. Turner: 'The first thing that caught my attention was the situation of the little town of St. Mary's, which so much resembled that of Reikevik that I could hardly help fancying for some time that I was in Iceland . . . nor is the surrounding country so much unlike as you would perhaps expect, for except where there are enclosures of stone the surface is equally barren.' He found mosses and lichens to be far from luxuriant in the islands, and at the time parched almost to a cinder, there having been no rain for many weeks.

In the same year he had two interesting visitors at Halesworth; one was his old friend Jorgen Jorgensen, who in a record of his life, printed in Tasmania[2], speaks of the hearty welcome he received, adding: 'Availing myself of the quiet retirement of this country residence I shut myself up and wrote an account of the Icelandic Revolution, in which I introduced various anecdotes of Scandinavian history. I presented it to Sir Joseph Banks.'

The other visitor was Dr. Thomas Taylor[3] of Dunkerron, Kerry, an excellent Irish muscologist, who spent three weeks with him over his own and Turner's herbarium. The latter was lent for the purpose, and was of especial interest as containing the types of the 'Muscologia Hibernica.' In a letter to Mr. Turner, Taylor is described as having been born in India, and up to his seventh year knowing no language but Hindustani; he was then shipped to Ireland in a vessel where

[1] Joseph Hooker, junr., died in 1815 of consumption, for which the treatment in vogue then, and for many years afterwards, was 'powerful medicines and abstinence from nourishing food.' He was an excellent British entomologist. His collection of insects was purchased by the British Museum: my father's is now in the Norwich Museum.

[2] Ross's Hobart Town Almanack and Van Diemen's Land Annual for 1835, p. 138. The article is anonymous, entitled 'A Shred of Autobiography.'

[3] Died at Dunkerron, 1848. He was joint author of the Muscologia Britannica.

nothing but Portuguese was spoken, and on landing sent to a school at Cork where French alone was heard. It was thus comparatively late in life that he acquired English, and this with an Irish accent.

Early in 1814 my father accompanied Mr. and Mrs. Turner and family on a visit to Paris, then in the occupation of the Allies. There, at 'The Institute,' he made the acquaintance of the principal botanists resident in, or on visits to the city—Antoine Laurent de Jussieu, Desfontaines, Lamarck, Mirbel, Bory de St. Vincent[1], Thouin, and others. Leaving the party in Paris he spent the remainder of the year botanizing and seeing botanists, sketching and sight-seeing in the south of France, spending some days with de Candolle at Montpellier, and in Piedmont, Switzerland, and Lombardy. Returning to Paris early in 1815 he was introduced to Humboldt, who engaged him to publish a cryptogamic volume of his 'Plantae Equinoctiales.' This intention had to be abandoned owing to the publisher's refusal to continue that work. After much subsequent correspondence with Humboldt, that led to nothing, my father commenced the publication on his own account, and produced in 1816 the first part of a work entitled 'Plantae Cryptogamicae, quae in plaga orbis novi Aequinoctialis colligerunt Alex^{r}. von Humboldt et Aimat Bonpland.' It is a very thin quarto with four plates of species drawn by the author, and exquisitely etched by Edwards. The expense was great and the return nil; the work was therefore abandoned, and of the remaining Musci and Hepaticae many were included in the author's less expensive 'Musci Exotici.'

On June 12, 1815, my father married Maria Sarah, eldest daughter of Dawson Turner, and immediately started on a long wedding tour to the Lake District and to Ireland, which latter country the pair traversed in almost every direction, making sketches of scenery and ancient buildings; thence they went to Scotland on a visit to Mr. Lyell[2] at

[1] Of the above, only two were alive to welcome me when I visited Paris in 1845; Mirbel died in 1854, Bory in 1846.

[2] Father of Sir Charles Lyell, translator of Dante. I have been unable to

Kinnordy in Forfarshire, with whom a close intimacy and correspondence on Hepaticae had long existed. Returning they passed through Manchester for the purpose of seeing Mr. Hobson [1], a packer in a warehouse, who with only the works of Withering, Hudson, and the 'Muscologia Hibernica' had acquired a critical knowledge of British mosses that surprised his visitor, who says of him: 'I never saw a man possessed of more enthusiasm than this poor fellow.'

Taking up his residence in Halesworth, my father was for the next four years seldom long without interesting and often distinguished botanical visitors. Early in 1816 he had staying with him, preparing to accompany Earl Amherst as medical attendant in his embassy to China, Dr. Clarke Abel [2], a young Norwich friend, whom he had recommended to Sir Joseph Banks for that appointment. Dr. Abel returned in 1817 and again stayed with my father at Halesworth, writing up his journal for publication and naming his plants.

In the same year M. de Candolle spent some days with him, of which the following account is given in the writer's own words [3]:—

'J'allai par les voitures publiques d'abord à Halesworth où demeurait M. Hooker. Il me reçut avec beaucoup d'amitié

discover the beginning of my father's intimacy with Mr. Lyell. It commenced when the latter lived at Bartley Lodge, in the New Forest, which he diligently explored for Hepaticae. In the introduction to the British Jungermanniae Mr. Lyell is mentioned as having suggested alterations in the arrangement of the species adopted by Lamarck and de Candolle in the Flore Française. He was elected F.L.S. in 1813.

[1] Edward Hobson, who died in 1830, was the author of two volumes 8vo of Specimens of British Mosses.

[2] Clarke Abel, M.D., F.L.S., had practised for a short time as surgeon in Norwich, when his devotion to natural history led him to seek employment abroad. After returning from China he entered the service of the East India Company and went to Calcutta, where he won the regard of Dr. Roxburgh at the Botanical Gardens. He died at Cawnpore in 1826. His description of the tame orang-outang in the Asiatic Researches is classical, as are his works on the wild dog of the Himalaya and the crocodile of the Ganges. His Narrative of a Journey in the Interior of China (London, 1818) gives an account of the misfortunes of the embassy.

[3] Mémoires et Souvenirs d' Augustin-Pyramus de Candolle, écrits par lui-même et publiés par son fils, p. 272. Genève, 1862.

et je logais quelques jours avec lui. Sa femme, qui était aussi distinguée par la figure et par l'esprit, me reçut également d'une manière très amicale. Nous passions nos journées ensemble à causer, surtout de botanique, à voir son herbier et les plantes qu'il cultivait dans son petit jardin. J'y fis connaissance avec Lindley, alors jeune élève de Hooker, et qui depuis est l'un des premiers botanistes de l'Angleterre. Madame Hooker est fille de M. Dawson Turner, botaniste, connu par un bel ouvrage sur les *Fucus*. Elle m'engagea à aller à Yarmouth voir son père, et je fus, en effet, reçu avec la plus franche hospitalité. Madame Turner était une mère de famille très distinguée et elle dessinait assez bien et gravait à l'eau forte. Mon portrait a été gravé par elle.'

As alluded to by M. de Candolle, Lindley, then a youth of eighteen, was at the same time with himself a guest of my father. He was the son of a well-known nurseryman of Catton, near Norwich, and had shown such zeal and ability as a local botanist that with a view of encouraging him in its pursuit he was invited to Halesworth [1], and to occupy himself there with translating Richard's 'Analyse des Fruits.' This he did, introducing the author's latest corrections, and illustrating his translation with plates and original observations [2]. In the following year my father took Lindley to Sir Joseph Banks, who offered him temporary employment in his herbarium, and introduced him to Mr. Cattley, a wealthy merchant devoted to horticulture, who was desirous of having his rare plants handsomely illustrated [3]; and this again led eventually to the assistant secretaryship of the Horticultural Society of

[1] On this occasion Lindley was looking forward to employment as a botanical collector abroad, which led to an amusing incident. The housekeeper at Halesworth finding that his bed was never occupied, after a vain search for a reason, reported the fact. His distressed host had to ask for an explanation, which was simply that his guest was inuring himself to the hardships of a collector's calling by sleeping on hard boards! Dr. Lindley died in 1865, three months after my father.

[2] Published in London, under the title of Observations on the Structure of Fruits and Seeds, in 1819, pp. 100, and six plates; dedicated to W. J. Hooker.

[3] The result was the publication of the Collectanea Botanica, a folio with forty-one coloured plates. London, 1821.

London, which Lindley occupied till 1858. In the same year he was visited by Professor C. Martens of Bremen, an enthusiastic algologist, and father of Professor Martens who accompanied the Russian Captain Lutke on his voyage to Behring Sea, where he made valuable observations and collections of the wonderful Algae of that sea; and later received a second visit from Dr. Taylor, who was engaged with his host on the 'Muscologia Britannica,' published in 1818 with twenty-eight plates illustrating 269 species and three tables of genera with thirty-two species. This work had taken in all eight years of preparation, nearly every species having been collected by one or both authors. The number described is 269 as against 227 enumerated by de Candolle for France, including the Pyrenees and Alps. The number of synonyms is about 470. A second edition hereafter to be notified appeared in 1827.

The 'British Jungermanniae,' the most beautiful of all my father's works, in point of the drawing, analyses, and engraving of the plates, was concluded in 1816. It had occupied him for about ten years, and was the first work of any magnitude which he projected. It appeared in parts, in both a quarto and a folio form, with eighty-eight plates engraved by Edwards, illustrating 197 species.

In the same year he commenced working for the new edition, by G. Graves, of Curtis's 'Flora Londinensis,' a sumptuous work, the parts of which appeared at long intervals from 1819 to 1828. Its perplexing issue will be described later on.

1817 is one of the very few years of his life in which he published scarcely anything. The exception was an account of the very remarkable European moss named after his friend, *Tayloria splachnoides*, in 'Brand's Journal of Science and Art,' No. III, p. 144, and 'Musci Exotici,' tab. 173. Of a visit to London in August of this year, he writes: 'I met at Spring Grove (Sir Joseph Banks's) Abel, Brown, Leach, and a Mr. Manning of Diss, who passed many years among the Chinese endeavouring to get access into the interior, though he failed; though he tells me he saw much of Thibet.' Mr. Manning is, to this day, the only Englishman who ever entered the sacred city of Lhassa. What is more remarkable is, that his journal was lost to geographers till Sir Clements Markham happily found it in the possession of a cousin of his own in Norfolk. See 'Narratives of the Mission of G. Bogle to Thibet and of the Journey of T. Manning to Lhassa,' ed. 2, 1879, by Sir C. Markham, a book full of curious information.

In 1818 my father had the pleasure of receiving at Halesworth Robert Brown, Dr. Burchell [1], Mr. Lyell of Kinnordy, and Dr. Boott [2] of Boston, Mass. The first volume of the 'Musci Exotici' appeared in this year, the second in 1820, in both octavo and quarto forms. One of the objects of the work was to illustrate Humboldt's and Bonpland's discoveries, of which thirty-five are figured; but the collections of Menzies during Captain Vancouver's voyage (1790–5), in New Zealand and North-West America especially, more than doubled that number. Other important contributions came from Burchell's Cape travels, Buchanan-Hamilton's Nepalese, and Brown's Australian. Altogether 176 species are figured, etched by Edwards, from coloured drawings by the author.

My father's Halesworth life was now drawing to a close: the brewery business, as might have been expected under the management of an enthusiastic naturalist and author, had proved unsatisfactory, and some of his investments were disappointing. Personally his ménage was entirely inexpensive

[1] Dr. W. Burchell, D.C.L., F.L.S., was a great traveller. Embarking in 1804 on a voyage to the Cape for botanical purposes chiefly, he spent five years en route in St. Helena, then five in South Africa, penetrating into the Transvaal before the immigration of the Boers. He then went to Brazil and travelled for four years in the interior. On both journeys he made enormous collections of plants, estimated at 15,000 species, and many views of scenery, for he was a beautiful artist. But except two quarto volumes of travels in Africa and descriptions of three new African animals, he published nothing, and he shut himself up in his museum at Fulham, where I visited him about 1860. On his death in 1863 his herbarium was presented by his sister to the Royal Gardens, Kew.

[2] Dr. Boott, M.D., F.L.S., secretary and treasurer of the Linnean Society, resided in London and devoted his life to the illustration of the genus *Carex*, upon which he published four folio volumes with plates. He died in 1863. His collections and all the drawings he had made of the genus were presented to the Royal Gardens, Kew, by his widow.

and simple, and this was so throughout his life; but his lavish expenditure on his own unremunerative publications, and on the purchase and beautiful binding of expensive entomological, ornithological, and especially botanical and even archaeological and artistic works, had crippled his resources, and he had now a wife and family of four to provide for. Under these circumstances he wrote to his friend Sir Joseph Banks, requesting that he might be informed, should he hear of any opportunity of applying his botanical knowledge to the improvement of his income. Sir Joseph promptly answered, that the Professorship of Botany was vacant in the University of Glasgow, and that he was ready to use his influence to obtain it for him should he desire to become a candidate. My father answered favourably, and at once left for Spring Grove, where he was hospitably received by Sir Joseph[1], who told him that the emoluments of the Chair, though small, would certainly increase; that it was freed from all medical duties[2]; that a really noble botanical garden had been formed at Glasgow, to which the University had given £2,000 and the city £3,000, and towards the development of which he could assure him that Kew would place all its resources.

[1] The securing this professorship for my father was probably the last of the good deeds of this truly noble soldier of science. He died in the following June (19) aged 77.

[2] The Chair had been held conjointly with that of medicine by Dr. R. Graham, who was now transferred to Edinburgh. The Edinburgh Chair, as was that of Glasgow, had been first offered to Robert Brown, who declined both, on the score of his obligations to the aged Sir Joseph Banks, whose librarian he was. Sir James Smith had been a candidate for the Edinburgh Chair.

CHAPTER II.

GLASGOW, 1820-1840.

EARLY in February, 1820, my father was appointed by the Crown to the Chair of Botany in Glasgow, and having dispatched his library, herbarium, and household effects to London, to be thence sent by smack to Leith, and on to Glasgow by canal, he severed his connexion with Halesworth and the brewery. In May he presented himself before the Senate of the University, who gave him a flattering reception, read his inaugural thesis[1] (the Latinity of which, thanks to his classical father-in-law, was highly praised), and was duly installed, with the welcome addition of having the honour of LL.D. conferred upon him.

Meanwhile the preparation for his course of botanical teaching, which commenced in May, had been for three months a grave anxiety. He had never taught, lectured, or even heard a course of lectures, and some important branches of the science he was called upon to profess were new to him. Such especially was the anatomy of plants, of which he writes: 'It is a subject to which I have never attended, and authors are so much at variance as to their opinions, and on facts too, that I really do not know whom to follow. Knight in every one of his papers contradicts what he himself asserted in former ones, and has got handsomely lashed for it in the second number of the 'British Review'; as has Sir James Smith, for adopting his theories and for

[1] Of this thesis I find no copy amongst my father's papers; and in answer to a request that the records of the University might be searched for it, I am informed that it does not exist there. It was entitled De Laudibus Botanicis.

giving him the highest praise for his perspicuity. I have written for Kieser's work[1] on the subject, which Brown says is the best. Mirbel has seen what nobody else can; so nobody contradicts him, though many won't believe him.'

Before enlarging on my father's success as a lecturer, I may premise that the teaching of botany in the first quarter of the last century was very different from that which now prevails. It was regarded as ancillary to that of Materia Medica, and as a means of enabling the practitioner to recognize the plants used in medicine when there might be no druggist to appeal to. Furthermore, it was required by the principal examining bodies for medical degrees or licences, that the candidate should have attended a course of lectures delivered in a botanical garden registered for the purpose; and in these gardens the plants were invariably arranged according to the Linnean[2] system, which consequently had to be taught. The latter was, however, with the new Glasgow Professor a secondary consideration, his primary aim being to open the eyes and minds of his pupils to the principles upon which plants were classified, and their distribution and uses, which was as much, he thought, as could be comprehended in a course of sixty lectures by young men who did not even know the elements of biology, and had not been exercised in using their eyes, hands, and brain in unison in the examination of a plant or animal. The course was opened by a few introductory lectures on the history of botany and general character of plant life. As a rule the first half of each hour was occupied with lecturing on organography, morphology, and classification, and the second half with the analysis in the class-room of specimens supplied to the pupils, the most studious of whom took these home for further examination. An interesting event in these half-hours was the Professor frequently calling upon such students as volunteered for being examined, to demonstrate the structure of a plant or fruit placed in the

[1] Grundzüge der Anatomie der Pflanzen. Jena, 1815.

[2] Some of the students of my father's first year's course remonstrated against his introducing the Natural System into his teaching.

hands of the whole class for this purpose. Throughout the course my father's artistic powers were exercised with chalk and the blackboard ; and he gradually accumulated a magnificent series of folio coloured drawings, especially of medicinal plants, which were suspended in the class-room as occasion required. I well remember the murmur, and even louder expression of applause with which he was greeted on taking the Chair, when the number or interest of these pictures was conspicuous. Before his second year's class had assembled he had published the 'Flora Scotica' for its use, and an oblong folio of lithographed illustrations of the organs of plants by his own pencil, with twenty-four plates and 327 figures, a copy of which was placed before every two students [1]. During the course three botanizing excursions were taken, two in the neighbourhood of Glasgow, and one towards the end of June, of five or six days' duration, to the Western Highlands, usually to the Breadalbane range. This latter was eagerly anticipated by a contingent of ten to thirty students, amongst whom were frequent accessions of botanists from Edinburgh and England. Further to stimulate their zeal, he habitually invited the more industrious students to breakfast with him after the class (which was from 8 to 9 a.m.), when he would show them books, and give them from his store of duplicates, specimens of rare British plants. To conclude this episode of his life, it must be recorded, that his success as a lecturer was phenomenal ; his tall figure, commanding presence, flexible features, good voice, eloquent delivery, and urbane manners are vouched for in every obituary notice of him. His lectures were often attended by gentlemen of the city, and even by officers from the barracks three miles distant. The students of his first year's course presented him with a handsome silver vasculum, chased with a design taken from the moss, *Hookeria lucens*, and those of the second year with a richly bound copy in ten volumes of Scott's Poetical Works.

[1] A second series in quarto with twenty-six plates, comprising 395 illustrations, by Fitch, for class use, was published in 1837.

During the twenty years of my father's Glasgow residence his life was one of continuous but congenial labour. For the first fifteen years or so he gave only one course of lectures, from May 6 to the middle of July, in the Botanical Gardens; but towards the end of his professorship a winter course was given in the College buildings. These and the examinations in botany for degrees were his only professional duties; the rest of his time was devoted to his botanical studies, drawings, and publications, the increase and keep of his herbarium, and rapidly accumulating botanical correspondence. Except for short visits to London, Yarmouth, or the Highlands, botanizing with Greville or Arnott, and once to Paris, he rarely left home. He was at his desk with pen or pencil by 8 a.m., and never left it much before midnight. The late summer and autumn weeks were frequently passed with his family at watering-places on the Clyde, usually at Helensburgh, where he enjoyed the society of two neighbours of scientific tastes and culture, James Smith, Esq., F.R.S., of Jordan Hill[1], and Lord John Campbell, afterwards Duke of Argyll, father of the late Duke, who inherited his parent's scientific tastes. In 1837 he purchased a cottage with an acre of ground, 'Inver-eck[2],' near Kilmun, on the Holy Loch; a lovely spot where he could indulge his fondness for gardening. In the touring season he received many English and foreign friends, who took Glasgow on their route for the Highlands, both to visit him and to avail themselves of his experience of roads, conveyances, and accommodation.

My father's reputation as one of the foremost botanists in this country was confirmed by his success in the Glasgow Chair, and rapidly rose as his successive publications appeared. Very soon he had but one compeer in Great Britain, Dr. Lindley, for Robert Brown towered above both as 'Botanicorum facile princeps.' It was a happy augury for

[1] Eminent as a geologist, and as author of The Voyage and Shipwreck of St. Paul.

[2] The site of the cottage is now occupied by a castellated mansion in the Scottish style of architecture.

the progress of the science which both worshipped with single-minded zeal, that Lindley and my father were regarded as meriting equal recognition as scientific botanists and indefatigable labourers throughout forty-five years of their active lives, and that they should have been fast friends till death, within three months of one another[1].

As his own reputation advanced so did that of his herbarium and library, which before he had been ten years in Glasgow were reckoned as amongst the richest private ones in Europe[2]. This was due to his active correspondence,

[1] The following admirable summary of the life-works of my father and Lindley respectively, is extracted from the Proceedings of the American Academy of Arts and Sciences, May 29, 1866:—'The names of Hooker and Lindley, which stood side by side in our botanical section, are naturally associated as those of the two most eminent botanists in Great Britain—also by the parallel course, and near coincidence in the close of their lives. Born in the same neighbourhood, in youth receiving their education at the same school, and early drawn together by similar predilections, they both devoted themselves with singular energy and perseverance to their chosen pursuit; exerted for many years, although in somewhat different ways, a paramount influence upon the advancement of botanical science; and died near together in place and time—the elder at Kew, on August 13 last, at the age of eighty years; the younger at Turnham Green, on the first of the ensuing November, at the age of sixty-seven years. For a long time they were the two most distinguished teachers in Great Britain, one at a northern, the other at the metropolitan University. They severally conducted two of the principal serial works by which botany contributes to floriculture; and they developed into highest usefulness the two great establishments, the Royal Gardens at Kew, and the Horticultural Society of London. Both wrote and published largely—Hooker only upon descriptive botany, in which he greatly excelled, while Lindley traversed a wider field, and grappled with abstruser problems in every department of the science, always with confidence and facility, but not with unvarying success.'

[2] The following testimony to the value of the herbarium is an extract from an essay on European Herbaria by Asa Gray, written in 1841, and published in the American Journal of Science and Arts, xl. 1 (see also Scientific Papers of Asa Gray,' ii. 13): 'The herbarium of Sir William J. Hooker, at Glasgow, is not only the largest and most valuable collection in the world, in possession of a private individual, but it also comprises the richest collection of North American plants in Europe. Here we find nearly complete sets of plants collected in the Arctic voyages of discovery, the overland journeys of Franklin to the polar sea, the collections of Drummond and Douglas in the Rocky Mountains, Oregon, and California, as well as those of Professor Scouler, Mr. Tolmie, Dr. Gairdner, and numerous other officers of the Hudson's Bay Company, from almost every part of the vast territory embraced in their operations from one side of the continent to the other. By an active and prolonged correspondence with nearly all the botanists and lovers of plants in the United States and Canada, as well as by the collections

judicious purchases, the contributions of his former pupils, especially from abroad, to his methodical habits, and to the welcome he gave to all botanists who desired to consult his collections. For the operation of mounting specimens, &c., he employed aids, of whom I remember two; the first, in about 1827, I think, was a native of Dundee, a keen algologist, James Chalmers by name, who prepared fasciculi of named Algae, in quarto form [1], in the disposal of which my father aided him. The other was Dr. J. Klotzsch, who spent some years as curator of the herbarium. Klotzsch was an excellent fellow, a devoted mycologist, and whilst at Glasgow would study no other branch of botany than fungi. During the summer and autumn months he frequently rose at 4 a.m. and made a long excursion collecting in the environs of the city. On these occasions his appearance excited great curiosity; he was short and stout, wore a green doublet and German peaked cap, his long hair flowed over his shoulders, a huge tin vasculum was strapped to his back, he carried a stout staff with a pickaxe head, and his English was very German. Meeting the rough factory hands and miners on their way to work he was often hustled and even assailed, when he defended himself with this weapon, and, being quick of temper, on one occasion felled with it a too rash tormentor. Klotzsch was the founder of the mycologic portion of the herbarium. Returning to Berlin, he took up the study of flowering plants, acquired distinction as a botanist, and became eventually Keeper of the Royal Herbarium, Berlin. The only other aids my father had in Glasgow were my mother, as amanuensis, and myself; for having been attracted to

of travellers, this herbarium is rendered unusually rich in the botany of this country; while Drummond's Texan collections, and many contributions from Dr. Nuttall and others, very fully represent the flora of our southern and western confines. That these valuable materials have not been buried, or suffered to accumulate to no purpose or advantage to science, the pages of the Flora Boreali-Americana, the Botanical Magazine, the Botanical Miscellany, the Journal of Botany, the Icones Plantarum, and other works of this industrious botanist, abundantly testify; and no single herbarium will afford the student of North American botany such extensive aid as that of Sir William Hooker.'

[1] Algae Scoticae. See Hook. Journ. Bot., i. 158.

botany from my childhood, much of my spare school and college time was devoted to the Herbarium.

In 1820 there were few botanists in Scotland to welcome the newly-appointed Professor, and of these only two were known to him personally, his old friend Mr. Charles Lyell[1], of Kinnordy, in Forfarshire, who had, however, abandoned the study of *Hepaticae* for that of Dante; and the Rev. Dr. Stuart, of Luss, with whom he had botanized during two of his Highland tours (see p. xiii). Others were Dr. Robert Graham, his predecessor in the Glasgow Chair, then holding that of Edinburgh; with him there was maintained a close correspondence till his death in 1845, chiefly concerning plants flowering in the Edinburgh Botanical Garden, many of which were figured in the 'Botanical Magazine'; Dr. R. K. Greville, LL.D., of Edinburgh, with whom he had corresponded when in Suffolk on the structure of *Buxbaumia aphylla*, and with whom, as associate, he published the 'Icones Filicum' and many papers on mosses and ferns; he died the year after my father, in 1866; Dr. Hopkirk, LL D., F.L.S., of Glasgow, author of the 'Flora Glottiana' and 'Flora Anomoia,' who had taken the leading part in the formation of the Glasgow Botanical Garden; Captain Dugald Carmichael, F.L.S., of Appin, Argyleshire, who had been a brother medical officer with Robert Brown in a fencible regiment stationed in Ireland; and Dr. F. Buchanan-Hamilton, F.R.S., F.L.S.[2], of the Indian Medical Service, who after a long and active career in India, including for a short period the superintendence of the Botanical Garden of Calcutta, had succeeded to an estate near Callender, where he died in 1829. These had all been correspondents except the now-forgotten

[1] Mr. Lyell died in 1849; retaining to the last his interest in botany, corresponding with my father on his publications, and responding liberally to the calls for aid and counsel from struggling botanical workers, and their widows and families.

[2] Author of 'An Account of the Kingdom of Nepal,' of 'A Journey from Madras through Mysore, Canara, Malabar, &c.,' and 'A Commentary on Rheede's Hortus Malabaricus' (Trans. Linn. Soc., vols. xiii, xv, xvi). He was the earliest explorer of the Flora of Nepal, and, after Rheede (in 1670), of Malabar. His very large collections were distributed by Wallich.

soldier-botanist Carmichael, of whom he had heard when staying with Sir Joseph Banks in 1820 from R. Brown, as a man who had visited the all but unknown and inaccessible island of Tristan d'Acunha in the South Atlantic, and had left no branch of its natural history unexplored. Captain Carmichael, then retired, was living in seclusion in a farm of his own at Appin, in Argyleshire, devoting his whole energies to investigating the cryptogamic flora, especially the Algae and Fungi, of his vicinity. Ten years subsequently my father published in the 'Botanical Miscellany' (ii, pp. 1, 258; iii, p. 23) a very interesting memoir of Carmichael, written by his friend, the Rev. Colin Smith, of Inverary, giving a full account of his military services, first as a medical officer, and latterly as a lieutenant and captain in his regiment, which was actively employed under Sir David Baird at the taking of Cape Town. The memoir gives long extracts from his journals on the botany, zoology, and physical geography of the countries around Cape Town and Algoa Bay, and of the islands of Mauritius, Bourbon, and Tristan d'Acunha, in respect of which one cannot but admire his powers of observation, and wonder how under the obstacles and discouragements of a soldier's life in those days he obtained the thorough scientific knowledge he displays, of the botany especially, of the several countries he visited. Of these latter Tristan d'Acunha was virgin soil, and of its natural history little is as yet known beyond what he recorded. The occasion of his visiting it was, that being at the Cape when orders were sent to take possession of it (as an eye over our prisoner Napoleon in St. Helena[1]), he obtained leave to accompany the expedition. This enabled him to spend between six and seven months in the island, which he devoted to its exploration. The result is a paper entitled 'Some Account of the Island of Tristan d'Acunha, and its Natural Productions[2],' by Captain Dugald

[1] The knowledge of geography possessed by the War Office of those days must have been rudimentary.

[2] Transactions of the Linnean Society of London, vol. xii, p. 483. This important paper is overlooked in the otherwise very full history of Tristan d'Acunha given in The Narrative of the Cruise of the *Challenger* (vol. i, p. 241).

Carmichael, which contains amongst other matters a complete account of its flowering plants and ferns.

Soon after his arrival at Glasgow my father had a visit from Captain Carmichael, bringing with him a collection of the mosses of Appin. Of him he writes[1]: 'It was impossible not to be struck with the varied knowledge he possessed, for though in botany he took the greatest delight, yet with almost every subject, and especially such as bore any relation to his extensive travels, his mind was richly stored. It was in examining these minute productions (Fungi and Algae) that he spent almost the whole of his life after his retirement from active service. And though his attention was wholly confined to the parish in which he lived, he was so eminently successful that among the Fungi alone he detected more species than had been before described as native of the whole of Scotland. My last interview with him was in the summer of 1826, when I invited him to join an excursion to the Western Islands with the students of my class. He met us in our vessel immediately opposite his residence, when we proceeded to Mull and Skye; thence returning through the Sound of Mull we visited Fort William, Ben Nevis, and the majestic scenery of Glencoe.' Captain Carmichael died in the following year. The results of his labour in Scotland appear in Part II of the 4th edition of 'The British Flora,' devoted to Algae and Fungi. Two manuscript volumes in 4to, 'Algae Appinenses' and 'Cryptogamiae Appinenses,' preserved in the Library at Kew, testify to his knowledge and skill as a botanist, microscopist, and artist.

Very soon after the settlement of the herbarium and library in Glasgow botanists from all parts of Europe flocked to it, amongst whom the following eight made the most frequent and longest sojourns, some of them becoming collaborators with the owner: R. K. Greville, G. Bentham, Sir J. Richardson, G. A. Walker-Arnott, W. Wilson, the Rev. M. J. Berkeley, H. C. Watson, and W. H. Harvey. Mr. Bentham's first visit was in 1823, from which occasion he dated his permanent

[1] Botanical Miscellany, vol. ii, p. 4.

adhesion to botany as an occupation for life. The next (in 1823) was Dr. (afterwards Sir John) Richardson, R.N., the companion of Franklin in his Arctic expeditions, through whom my father was made known to the Lords of the Admiralty, the Directors of the Hudson's Bay Company, and the chiefs of the Colonial Office, thus becoming the recipient of many herbaria made by the officers of these departments, and the author of works published under their authority. It further led to his being asked to recommend young medical men fond of natural history, from amongst his pupils especially, to embark in their services abroad.

In 1825 he first met Mr. G. A. Walker-Arnott, of Arlary, a member of the Scottish Bar, then living in Edinburgh. I think I am correct in saying that their meeting took place in Paris, when my father being taken ill was kindly attended to by this fellow-countryman, a stranger to him, who was staying in the same hotel. Mr. Arnott must then have been on his way to join Mr. Bentham in his exploration of the botany of the Pyrenees [1]. He was a mathematician of considerable attainments [2], and had published an important paper on the 'Classification of Mosses [3].' With him as collaborator were published the 'Botany of Beechey's Voyage,' 'Contributions towards the Flora of South America and the Pacific Islands,' and the sixth and seventh editions of 'The British Flora.' As Dr. Arnott, LL.D., he succeeded Dr. Balfour in the botanical Chair of Glasgow, and died in 1868.

In 1827 my father's correspondence commenced with W. Wilson, of Warrington, who paid many visits, sedulously studying the mosses of the herbarium, and exploring the Highland mountains, sometimes joining in the botanical class excursions. Latterly he volunteered a revision of the whole collection of Musci in the herbarium, thereby adding very

[1] This resulted in the publication by Mr. Bentham of his Catalogue des plantes indigènes des Pyrénées et du Bas Languedoc. Paris, 1826.

[2] Author of two papers, 'On the Solutions of Experimental Equations,' and 'A Comparison between the Chords of Arcs employed by Ptolemy, and those now in use' (Tilloch's Phil. Mag., 1817 and 1818).

[3] Mém. Soc. Hist. Nat. Paris, 1825.

largely to its value. He co-operated in the publication of many papers on exotic mosses in my father's 'Journals of Botany,' and edited a greatly enlarged edition of the 'Muscologia Britannica' under the title of 'Bryologia Britannica.'

In 1828 my father first became acquainted with the Rev. M. J. Berkeley, of Kings Cliffe, Northamptonshire, the mycologist, who was then, I believe, on his way to visit Captain Carmichael in Appin. This led to a very intimate friendship, and repeated visits to West Park and Kew. Mr. Berkeley took the same interest in the Fungi of the herbarium as Mr. Wilson did in the Musci, and but for him this order of plants would never have attained its present pre-eminence; for his zeal induced my father to urge his correspondents in all parts of the world to collect fungi; with what success is shown by the richness of his herbarium, and the numerous papers on exotic genera and species of the order published by Mr. Berkeley in the botanical Journals, in the 'Transactions of the Linnean Society,' and many other works. Mr. Berkeley also contributed the volume on fungi to the third edition of Hooker's 'British Flora' (vol. v, p. ii of Smith's 'English Flora'), and, dying in 1889, he bequeathed his herbarium to Kew, together with the choice of his botanical library.

In 1830 Mr. Hewett Cottrell Watson, the most accomplished of British botanists, then resident in Edinburgh, requested permission to accompany the students of the botanical course on an excursion to the Breadalbane Mountains, for the purpose of ascertaining the altitudes affected by their plants. Thus commenced a very active and interesting correspondence between my father and this acute botanist, which led to the publication of many papers in the Journals conducted by the former, to the botanical expedition of the latter to the Azores, and indirectly to his valuable account of the Flora of that interesting archipelago[1] in Godman's 'Natural History of the Azores' (London, 1870).

In 1831 Mr. W. H. Harvey, of Limerick (afterwards Pro-

[1] London Journal of Botany, vols. ii, iii, and vi.

fessor of Botany in the Royal Dublin Society, and Keeper of the Herbarium, and eventually Professor of Botany in Trinity College, Dublin), introduced himself by letter with specimens from two new localities of a West Indian moss (*Hookeria laete-virens*) found nowhere in the eastern hemisphere but the south and west of Ireland. It was answered by an invitation to Glasgow, which resulted in an intimacy that amounted to his being regarded as a member of the family.

Mr. (afterwards Dr.) Harvey took the same interest in the Algae of the herbarium as Wilson did in the *Musci* and Berkeley in the Fungi, besides augmenting it largely by contributions from the splendid collections made during his voyages to North America, Ceylon, Australia, and the Pacific Ocean, for the sole purpose of investigating their marine Floras. Many illustrations of his classical works 'Nereis Boreali-Americana,' 'Nereis Australis,' 'Phycologia Britannica,' and 'Phycologia Australis,' were lithographed by himself under my father's roof at Glasgow, West Park, and Kew. He died the year after the latter[1], when staying with my mother at Torquay.

I must not close this brief notice of my father's activity in encouraging others without an allusion to the solicitude with which he fostered my own aspirations to become a traveller and a botanist; the interest he took in my ambitious projects; the energy with which he aided me in overcoming every obstacle thrown in my way, and prevailed on the higher powers to grant me facilities and the necessary funds; and, last but not least, the liberality with which he helped me whenever other resources were exhausted. In this connexion I refer especially to four crises in my scientific career:—my appointment to accompany Sir James Ross in the Antarctic Expedition in 1839 (for which he supplied all my scientific outfit); my (unsuccessful) candidature for the Professorship

[1] No fewer than ten of my father's botanical friends died within three years of his own end, including seven who were amongst his most intimate associates: Borrer, 1862; Boott, 1863; Lindley and Richardson in the same year (1865); Greville and Harvey, 1866; Arnott, 1868. The others were Burchell, 1863; Woods, 1864; Daubeny, 1867.

of Botany in Edinburgh University in 1845; my mission to India in 1847; and my appointment as Assistant Director of Kew in 1855. Add to these benefits, the legacy of his herbarium and library, and the truth of the saying 'one soweth, another reapeth' forcibly applies.

The works published by my father when in Glasgow are very numerous. A complete list of them, with details regarding the more important, will be given at the end of this sketch. They may be grouped under four headings—British Botany, American Botany, Miscellaneous Works, and Serials.

In the British Botany there was the 'Flora Scotica,' the new edition of Curtis's 'Flora Londinensis,' four editions of the 'British Flora,' and many contributions to a knowledge of British plants in the volumes of his botanical Journals.

The more important works on American Botany were the 'Flora Boreali-Americana'; Botanical Appendices to the Narratives of Sir E. Parry's three last voyages to the Polar Seas. There were also, in his botanical Journals, descriptions of T. Drummond's and of Geyers' United States and Oregon plants, and articles on the botany of Peru and Chili; and, in conjunction with Arnott, the 'Flora of South America and the Pacific Islands.' Also many American plants are described in the 'Botany of Beechey's Voyage' by himself and Dr. Arnott, and in his 'Icones Plantarum.'

Under Miscellaneous Works may be classed as most important Greville's and Hooker's 'Icones Filicum,' and the commencement of an 'Enumeration of all known Filices and Lycopodiaceae' by the same authors in the 'Botanical Miscellany'; the 'Botany of Captain Beechey's Voyage to Behring's Sea, the Pacific Ocean, and China' by himself and Arnott; the third edition of Woodville's 'Medical Botany'; the botanical articles in Murray's 'Encyclopedia of Geography,' and the first three volumes of the 'Icones Plantarum,' or figures and descriptions of new or rare and otherwise interesting plants contained in his herbarium, with 300 plates by W. Fitch[1].

[1] Walter Fitch, F.L.S., who by his artistic talents contributed so largely to

Of the Serial Works the first was the 'Exotic Flora,' inspired by the interest he took in the Glasgow Botanical Gardens. It was commenced in 1823 and concluded in 1827, with 232 coloured plates, from drawings mostly executed by himself, of exotic cultivated plants. It was followed in 1827 by his undertaking the authorship of the 'Botanical Magazine.' Of this work thirteen volumes were issued from Glasgow, the drawings for the first ten of which (about 720) were by his own pencil.

In the same year (1827), finding that his extensive correspondence with botanists and travellers abroad provided him with information of great value that might otherwise never see the light, and that his herbarium was at the same time teeming with plants unknown to science, my father formed the plan of himself editing a periodical for the diffusion amongst botanists of the information obtained from these sources. As a model he took Konig and Sims's 'Annals of Botany,' of which two volumes only had been published (London, 1805–6). He never stopped or stooped to calculate the time, worry, and cost that this undertaking would entail upon him, which occupied him for the next thirty years of his life; for he had throughout no assistant editor, and was dependent solely on my mother, and at intervals on myself when at home, for aid in proof-reading, &c. The heavy correspondence it entailed was conducted by himself alone.

Including the continuation of the series issued from Kew, these periodicals embrace twenty-eight volumes with 548 plates, of which seven volumes with 247 plates, the greater number of them drawn by himself, were issued from Glasgow. These were the 'Botanical Miscellany,' three volumes with 152 plates (1830–3), the 'Journal of Botany,' two volumes with 44 plates (1834 and 1840), and the 'Companion to the Botanical Magazine,' two volumes with 51 plates (1835–6).

the value of my father's works, was, when he entered the service of the latter (1834), a pattern-drawer in a calico-printing establishment in Glasgow, aged 18. His earliest work was for the Botanical Magazine and Icones Plantarum. He died at Kew, in 1892, in the receipt of a pension from the Crown (Civil List).

In the interval between the publication of the 'Companion to the Botanical Magazine' and the resumption of the 'Journal,' he undertook the editorship with Sir William Jardine and others of Taylor's 'Annals of Natural History,' which for three years (1837–40) was the recipient of much of his botanical matter; but the latter became too copious to be included in the numbers of the 'Annals,' and, the result proving otherwise embarrassing, that editorship was abandoned. After leaving Glasgow for Kew he resumed the 'Journal,' three volumes (1840–2) of which were followed by the 'London Journal of Botany,' seven volumes (1842–7), and that by the 'Journal of Botany and Kew Garden Miscellany,' nine volumes (1849–57).

Regarding the conduct of this series of journals under their different titles, it is impossible to overrate the value of the assistance and encouragement which he received throughout in contributions from fellow botanists at home and abroad; especially Arnott, Bentham, Berkeley, Harvey, W. Wilson, Hewett Watson, and Asa Gray: and this though, owing to the limited circulation of the volumes, the publishers (of whom there were consecutively seven) gave the contributors neither the work nor copies of their papers, except on payment. The editor contented himself with one copy, and gave his services, and in many cases the drawings on stone, gratuitously.

As a contribution to the history of botany during three decades of the nineteenth century these periodicals were unique; no period or subsequent decade of that century can show so rich a store of valuable botanical material. Amongst their most interesting contents are the letters from correspondents abroad—Jameson and Hall from Ecuador, Douglas from North-West America, T. Drummond from Canada and the United States, Spruce from the Amazon, Peru, and the Pyrenees, Purdie from Jamaica and New Grenada, Bromfield from the United States, Geyer from Oregon and the Rocky Mountains, Seemann from Panama and the North Pacific, Gardner from Brazil, Walker from Ceylon, Stocks from Sind and Beluchistan, A. Cunningham from New South Wales,

Fraser from Queensland, J. Drummond from South-West Australia, von Mueller from Victoria and tropical Australia, and many others. The articles headed 'Botanical Information' and 'Notices of Books' are full of instructive information written for the most part by himself.

Towards the end of his Glasgow life my father resumed a systematic study of ferns, which he had begun with Greville soon after his arrival there, the first result of which was the commencement of an 'Enumeration of all known Ferns,' published in the 'Botanical Miscellany.' The issue in parts of Hooker and Bauer's 'Genera of Ferns' was begun in 1838; it originated in his having been shown the beautiful analyses of many genera of the order by the veteran botanical artist Francis Bauer [1], who offered the loan of these for publication to my father; not that the order had in the meantime been neglected by him, as is proved by the numerous genera and species described and figured in his journals, in the 'Icones Plantarum' and other works, and by his publication of J. Smith's 'Genera of Ferns [2].' As I propose to give in an appendix to this sketch of his life a complete account of my father's works, I shall not further dwell here on those devoted to ferns.

[1] Francis Bauer, an Austrian by birth, came to England in 1788, and was through Sir Joseph Banks's influence attached to the Royal Gardens, Kew, with the title of Botanical Painter to the King. He resided in a cottage on Kew Green, where I visited him with my father in 1835, when he showed us the original daguerreotype plates of Niepce. He died at Kew in 1840, aged eighty-two, in the enjoyment of a pension left him by Banks. A handsome tablet in Kew Church records his career, and a fine oil painting of him hangs in the Kew Gardens Museum, No. 1. His published works are not numerous; the principal are, besides the Genera Filicum, Illustrations of Orchideous Plants, 20 plates, with a preface by Lindley; Strelitzia depicta, with 4 plates; and the plates (20) in Aiton's Delineations of Exotic Plants cultivated in the Royal Gardens of Kew, a huge folio (1796). His brother Ferdinand, equally celebrated as a botanical artist, accompanied Brown in that capacity on Flinders's survey of the coasts of Australia (1802–5).

[2] 'An Arrangement and Definition of the Genera of Ferns, with Observations on the Affinities of each Genus,' by J. Smith, Curator of the Royal Gardens, Kew (Journ. Bot., iv. 38–147; Kew Gard. Misc., i. 419, 659, ii. 378).

CHAPTER III.

West Park and Kew, 1841–1865.

During his occupation of the Professorship of Botany in Glasgow University my father, feeling keenly his severance from the scientific society of London, was always on the look-out for a congenial position there, even if of less emolument than that which he held. The Professorship of Botany in the newly created University College of London (then entitled London University) was pressed on him by Lord Brougham, but the possibility of an appointment to the Royal Botanic Gardens of Kew had for some years eclipsed all other prospects. Nor were his aspirations in this direction unreasonable, for over and above his botanical qualifications he had inherited a taste for cultivating plants, encouraged by ten years' experience in his own garden, greenhouse, and stove at Halesworth; he had twenty years' of good work in and for the Royal Botanic Gardens of Glasgow, and had been for thirteen years author of the 'Botanical Magazine,' a serial devoted to the illustration and description of cultivated plants. Added to this was the fact that Mr. Aiton, who as 'Gardener to Her Majesty' had controlled the Gardens of Kew since 1793, was approaching the age for retirement. Meanwhile the Kew Botanic Gardens, which for upwards of half a century had ranked as the richest in the world, had since the deaths, almost contemporaneously, of King George III and Sir Joseph Banks, been officially cold-shouldered, and had retrograded scientifically. Their early history is summarized in the official 'Guide-book to the Royal Gardens,' and need not be repeated here. The following is a *résumé* of the circumstances that led to their transference from the private property of the Sovereign to the nation as a scientific establishment under my father, who came forward as a candidate for their control on the first hint of a change in their management being contemplated.

Soon after the accession of Her late Majesty a revision of the royal household became necessary, and the question of retaining the Botanic Gardens at Kew as a royal appanage having to be considered, a Commission was appointed by Parliament to report upon them. The Commission, the chairman of which was Dr. Lindley, reported favourably on the whole, and concluded with the recommendation that they should be retained and extended, in the following words:—'The importance of Botanic Gardens has for centuries been recognized by the governments of civilized states, and at this time there is no European nation without such an establishment except England. The wealthiest and most civilized country in Europe offers the only European example of the want of one of the first proofs of wealth and civilization. There are many gardens in the British colonies and dependencies, as Calcutta, Bombay, Saharunpore, the Mauritius, Sydney, and Trinidad, costing many thousands a year: their utility is much diminished by the want of some system under which they can all be regulated and controlled. There is no unity of purpose among them; their objects are unsettled, their powers wasted from not receiving a proper direction; they afford no aid to each other, and, it is to be feared, but little to the countries where they are established; and yet they are capable of conferring very important benefits on commerce, and of conducing essentially to colonial prosperity. . . . A National Botanic Garden would be the centre around which all these lesser establishments should' be arranged; they should all be placed under the control of the chief of that garden, acting with him and through him with each other, recording constantly their proceedings, explaining their wants, receiving supplies, and aiding the mother country in everything useful in the vegetable kingdom; medicine, commerce, agriculture, horticulture, and many branches of manufacture would derive considerable advantage from the establishment of such a system. . . . From a garden of this kind Government could always obtain authentic and official information upon points connected with the establishment of new colonies: it would afford the plants required on these occasions, without its being necessary, as now, to apply to the officers of private establishments for advice and help. . . . Such a garden would be the great source of new and valuable plants to be introduced and dispersed through this country, and a powerful means of increasing the pleasures of those who already possess gardens; while, what is far more important, it would undoubtedly become an efficient instrument in refining the taste, increasing the knowledge, and augmenting the amount of rational pleasures of that important class of society, to provide for whose instruction is so great and wise an object of the present administration.'

Dr. Lindley's recommendations as embodied in the Report having become widely known, enthusiastic advocates of them soon made themselves heard, and a memorial urging their adoption, drawn up by the Linnean and Horticultural Societies and the University of London jointly, was addressed to the Government, and transmitted through the Treasury.

But to carry out such a scheme was not so simple a matter as at first sight appeared. There were many conflicting interests in high places to be consulted and conciliated during the three years' interval that elapsed between the sending of the Report to the Treasury and its presentation to Parliament. These were the Lord Steward (Earl of Surrey), under whose control the Royal Gardens were placed; the Commissioners of Woods and Forests, the Chancellor of the Exchequer, and Parliament itself. The initial difficulty arose from the position of the Botanic Gardens. These, though comparatively small, occupied a very important site in the royal demesne at Kew, by far the greater part of which latter, including a royal palace, were to be under any circumstances retained as such. It was not the Botanic Garden only that was wanted, but the attached Arboretum and space for indefinite extension, and hence inevitable interference with the amenities and privacy of the palace and its approaches. Even more formidable obstacles were the large expenditure which would have to be incurred on creating such an establishment as Dr. Lindley had outlined, and the best means of controlling it.

In the above-cited interval the fate of the Botanic Gardens was all but sealed, as the following extract taken from a little work, the author of which was a witness of the occurrence which he describes, proves:—'In the autumn of 1839 the Lord Steward, then Lord Surrey, who in virtue of his office had the whole management of the Royal Gardens, paid frequent visits to the Botanic Gardens, always accompanied by the superintendent of the kitchen garden, and carefully examined the greenhouses and pits; and it became known that it was his intention to convert them into vineries and pine stoves, and that the plants had been offered to the Horticultural Society for their garden at Chiswick, and also to the Royal Botanic Society for their garden at Regent's Park; but the offer was in both cases declined. The vinery scheme was, however, intended to be carried out, and on February 18, 1840, the kitchen gardener informed me that he had received instructions from Lord Surrey to take possession of the "Botany Bay House" and convert it as soon as possible into a vinery, and that the "Cape House" was to follow; and to enable him to do so he was to destroy the plants[1]. This becoming known to the public led to articles in the public journals condemning the scheme as being a disgrace to the nation. This had the desired effect, and Lord Surrey's scheme was abandoned[2].'

[1] The collections of Australian and S. African plants were unique. Some of the specimens were half a century old.

[2] Records of the Royal Botanic Gardens, Kew, by John Smith, ex-Curator. London, 1880. That the Royal Botanic Gardens maintained any position as a scientific establishment in the interval between the death of Sir Joseph Banks in 1820 and the appointment of the new Director in 1841, was wholly due to the unaided exertions of Mr. Smith, who, from being a foreman under Mr. Aiton, became Curator from 1841 till 1864 under my father. He kept up the correspondence with the colonial Gardens in the West Indies, S. Africa, and Australia, himself sowing the seeds and raising the plants that these contributed, and carefully recording their scientific names, habitats, and donors' names. For

Meanwhile unobtrusive but powerful influence was being exerted in favour of Dr. Lindley's recommendations by John, sixth Duke of Bedford, a nobleman distinguished for his devotion to botany, horticulture, and agriculture, who delighted in having the plants in his gardens at Woburn Abbey scientifically classified and named, as objects for his own gratification and study, and as materials for the production of botanical works of high scientific value, which he had printed and distributed at his own cost[1].

a notice of his life and labours see vol. ii, p. 429 of these 'Annals,' where it is stated that the characters of twenty new genera of ferns published in Hooker and Bauer's Genera of Ferns, are by him. But except *Ochropteris,* only sub-genera or sections of genera of ferns are there referred to.

[1] The services rendered to botany, horticulture, and agriculture by the sixth Duke of Bedford have been veiled through the suppression of his own personality in all he undertook for the encouragement of science and art. They were known well to few besides my father, whose botanical reputation the Duke recognized as early as 1817, and with whom he corresponded actively during his later years. Besides contributing liberally to the botanical missions of Schomburgk in Guiana, Purdie in New Grenada, Gardner in Brazil, Drummond in Florida and Texas, Tweedie in Argentina, Lippold in Madeira, and Cuming in Luzon, he established at Woburn a *Hortus gramineus*, an *Ericetum*, a *Salictum*, and a *Pinetum*, all on a scientific basis; and during his last illness he was forming a collection of Cacti, of which he had 450 species, and was contemplating an *Arboretum*. The following works were due to his munificence:—Hortus Gramineus Woburnianus, a folio volume of specimens, with an account of the results and experiments on the produce and nutritive qualities of different grasses and other plants used for the foods of the more valuable domestic animals; instituted by John Duke of Bedford, pointing out the kinds most profitable for permanent pasture, irrigated meadows, dry and upland pasture and alternate husbandry, with characters of the species and varieties, by G. Sinclair, gardener to his Grace.' Ed. i appeared in 1816; ii in 1825; iii in 1838.

Hortus Ericeus Woburnensis, a catalogue of 400 heaths, with four coloured plates, four of views, houses, and plans, and two of schemes of colour by Sir G. Hayter.

Salictum Woburnense, a catalogue of the willows in the Woburn collection, with coloured plates and descriptions by James Forbes, gardener to his Grace, roy. 8vo, 1829 (160 species and varieties are included).

Hortus Woburnensis, a descriptive catalogue of upwards of 6,000 plants cultivated at Woburn Abbey, with a view of the Abbey, twenty-six plates of plans of houses and beds, and brief descriptions of the species, by James Forbes, 1833.

Journal of a Horticultural Tour on the Continent, taken under orders of the Duke, in Hamburgh, Germany, Belgium, Bavaria, and France, by James Forbes, 8vo, 1837. It resulted in the acquisition of 600 new species to the Woburn garden.

Pinetum Woburnense, a catalogue of Coniferous plants in the Woburn collection,

Long before the issue of Dr. Lindley's Report to the Treasury the Duke, who knew the Gardens well, had entertained the hope that they might one day become the nucleus of a botanical establishment worthy of the nation; and in 1834, on an (unfounded) rumour of coming changes in their management being circulated, he warmly impressed upon his most influential friends my father's claims to be employed in them. When, therefore, in 1838, Dr. Lindley's Report came to his knowledge he wrote to my father expressing his satisfaction with it, entirely agreeing in its recommendations, and adding: 'though the outlay of restoring the Gardens to their original design and intention must of course be considerable, he should nevertheless think that no House of Commons would refuse a grant for an object so important in a national point of view.' In a letter dated Geneva, March 24, he writes: 'I look with hope and confident expectation to the prospect of seeing Kew Gardens and the whole of the surrounding demesne converted into a great national establishment, which may not only rival but be superior to the *Jardin des Plantes* at Paris. I have written to some influential friends to interest them in the subject.' And in another letter, alluding to his being solicited to patronize the establishment of a Garden in London, he says: 'I have uniformly refused, thinking that it would interfere with the more important plan of a great National Botanic Garden at Kew. I do not know if you are acquainted with the locale, but there is a large space behind the present Garden, now occupied as an unprofitable lawn and useless pasture, which is capable of being converted into a range of Gardens more useful as well as more ornamental than those of Paris.'

The Duke continued his efforts to induce the Government to give effect to the recommendations of the Report till within four days of his death[1]. This occurred on October 20, 1839,

systematically arranged and described by James Forbes, with sixty-seven coloured plates, roy. 8vo, 1839. In my father's collected correspondence there are 200 letters from the Duke almost exclusively on botanical subjects.

[1] His last letter to my father, written on October 16, the day before he was

after earnestly commending the measure to the care of his two sons, the seventh Duke, Francis, and Lord John (afterwards Earl) Russell, who was then in the ministry; and faithfully the sons carried out their father's wishes.

On March 3, 1840, reports being still in circulation that the Government intended to abolish the Botanic Gardens, the Earl of Aberdeen in the Upper House rose to inquire if such was the case, adding that he considered that establishment to constitute a part of the state and dignity of the Crown, which ought by no means to be alienated from it. He was answered by Viscount Duncannon that there was not only not the least intention to break up those Gardens, but there never had been such intention. To this Lord Aberdeen rejoined that he could assure the noble Viscount that an offer of the plants had been made to the Horticultural Society of London, and that the Society refused the offer, thinking it would be injurious to the public interest that the establishment should be broken up. Viscount Duncannon replied that though the care of the Gardens was not in his department, he had the authority of the Lord Steward for stating that no intention of breaking them up now existed.

It cannot but have been a source of regret with the Duke of Bedford's family and friends that he should not have lived to greet this, the dawn of the realization of his long-cherished wishes and hopes, together with the announcement which soon followed, that Her Majesty had graciously relinquished the Botanic Gardens and Arboretum of Kew, with the view of their being available for the public good.

On March 31, 1840, the Gardens, Pleasure Grounds, and Deer Park of Kew were (with the exception of about 20 acres surrounding a Swiss cottage) transferred from the Lord Steward's department to the Commissioners of Woods and Forests, the chief of whom was Lord Duncannon, now virtually pledged, by his answer to Lord Aberdeen, to maintain the Botanic Gardens. But that nobleman being firmly opposed

stricken with paralysis, ends with 'I have written to-day to Lord John to urge him strongly not to relax his efforts in pursuit of this grand object.'

to any enlargement of them and to any further expenditure upon them, all hopes of their forming the nucleus of an establishment worthy of the nation appeared, for the time at any rate, to be frustrated. The view taken by the Government of the expenditure and the responsibilities to be incurred in establishing the Gardens on a national footing, may be learned from a Report addressed to the Treasury by the Commissioners of Woods and Forests, signed by Lord Duncannon and other members, dated April 24, 1839. It points out that Parliament must find, in addition to the present annual expenditure on the Botanic Gardens, £20,000 for new works; and goes on to say, that though the services of the Board are available for the execution of the new works and supervision of the annual expenditure, neither it nor its officers can efficiently assist in the scientific management of the establishment and its adaptation to useful purposes; adding that such management and control would be most properly invested in trustees to be named by Her Majesty and to consist of persons holding high office in the State, and others at the head of institutions in the metropolis for education and science, as suggested in the Report of the Committee [1].

It is difficult to see how £20,000 could be profitably expended on new works in so confined an area as the Botanic Gardens then occupied.

After the Bedford family, Lord Monteagle, when, as Mr. Spring Rice, he was Chancellor of the Exchequer, was the most powerful advocate for the retention of the Botanic Gardens and for my father's being placed at their head; but at that particular time the national finances were in a straitened condition, and he could not propose a vote in the House of Commons for a Kew subsidy; nor could he influence Lord Duncannon in favour of the Botanic Gardens. Writing from Glasgow to Mr. Turner in December, 1840, my father says: 'From Lord Monteagle, indeed, I hear that the obstacles to my having an appointment at Kew are insurmountable.

[1] To which Report this refers does not appear. It is certainly not that of February, 1838, drawn up by Dr. Lindley, which contains no such recommendation.

My vexation, however, would have been greater than it is, had I not almost on the same day received a long and most kind letter from the Duke of Bedford, from which it does appear that I may now safely leave the matter in the hands of Lord John Russell; and I think I may infer from one or two parts of the Duke's letter, that however influential Lord Monteagle may have been when I first applied to him, when he was Chancellor of the Exchequer, he is now, being out of place, out of power. Lord Monteagle's letter was so decisive that I thought it right to thank him for all he had done in my behalf, and to close altogether the correspondence as bearing on Kew; and I have told the Duke that I should leave the matter in his brother's hands.'

The following letter from the Duke further explains the situation:—

'WOBURN, *December* 5, 1840.

. . . 'With respect to yourself and your own views and wishes, I do not like to be too sanguine or to hold out expectations that may not be realized, since we all know how much there is often between the cup and the lip, but I can give you some information which you will be pleased to hear. Lord Melbourne and my brother are here, and yesterday I had a full conversation with them and Lord Duncannon after breakfast on the subject in question. The result is that my brother, who is in truth your best friend, has desired Lord Duncannon to give him in writing a statement of the expenses that would be incurred to the public or to the Woods and Forests by your taking the place of Mr. Aiton at Kew. If that statement is satisfactory to him, which I have no doubt it will be, after what you stated to me in your letter of the first, he will himself prepare the matter to go to the House of Commons. I am sure from the decided manner in which he has taken it up that he will go through with it, if nothing unforeseen arises. In short, he will take it upon himself to make the proposition he is disposed to do in a way that justifies me in saying that he is your *best friend* in this matter; but I beg you to consider this as confidential.

'I have now got to the end of a long letter, written in great haste, but I trust I have answered all your points satisfactorily.

'Yours faithfully,

'BEDFORD.'

The continuation of the proceedings is best described in the following letter to Mr. Dawson Turner:—

'LONDON, *January* 24, 1841.

'I had received two letters from the Duke of Bedford, telling me how actively his brother Lord John was engaged in my interest in respect of Kew; and together with the last of them, one from Lord John, or rather from his private secretary, Lord E. Howard, written on the fourteenth, sent to me in Scotland, requesting that I would see Lord John in Downing Street the middle of this last week. I therefore felt it my duty to start immediately (from Jersey), and there was fortunately a vessel about to sail on Friday morning about 8 o'clock in which I embarked. We did not reach Southampton till nearly two on Saturday morning (yesterday), and the earliest train I could take was eight, and then I was obliged to leave my luggage in the Custom House, which I am sorry to say has not yet followed me. All dirty as I was I called in Downing Street, and had soon after an interview with Lord John, who seemed pleased with the promptitude with which I had come, and said he thought it much better that we should talk than write on this subject of Kew. He then again asked my opinion of Lindley's estimate[1], and for how much less expense the Garden could be carried on. I told him, knowing that economy was a great point with Lord Duncannon, that if, as I understood, £3,700 a year was now expended upon the Gardens, irrespective of Aiton's salary, I should feel well satisfied to conduct them with that income, feeling that with zeal and energy a great deal might be done which money could not buy. He asked what accommodation I should require on account of my herbarium and library (which he seemed duly to appreciate), which with that and my small salary was all the additional expense to be incurred. If, then he said, the £3,700 a year now paid from the Civil List did include Aiton's salary, he should have no hesitation in asking the House of Commons for £1,000 a year to cover that, and upon that ground the difficulties he thought might be removed. He further told me that Lord Duncannon had a serious objection to any additional ground being taken from the adjoining parks.

[1] This refers to a communication between Dr. Lindley and the Treasury on the subject of the cost of making Kew a botanical establishment worthy of the nation.

Then I said, "There are eighteen acres, let us see what can be done with them." "Now then," he added, "you had better see Lord Duncannon" (with whom, as with Lord Melbourne, he has had frequent communications). I told him I did not know him, and would be glad of a line from him. "Say I begged you might have an interview." I called, sent up my card with the above message, but the answer was, "His Lordship is very busy, and will be so busy on Monday and Tuesday that he cannot see you till Wednesday." This is always the kind of reception I have met with in attempting to see Lord Duncannon[1]. I am sure that with him, in reality the most influential man connected with Kew, there are obstacles that Lord Monteagle was justified in considering "insuperable." I believe more than ever that Lord Duncannon's great *desire* is to abolish the Gardens and save the expense to the Civil List. If he is determined on this I then think that Lord John will appeal to Parliament, for the Duke of Bedford was very explicit in assuring me that if his brother failed in one way he was prepared to try another.'

It was not till the following March that my father was officially informed that the Treasury had sanctioned his being appointed Director of the Botanic Gardens at Kew, with a salary of £300 and £200 allowance for the rent of a house. On the 26th of that month Mr. Aiton, under instructions from the Commissioners of Woods and Forests, transferred the Botanic Gardens and Arboretum to the new Director, reserving all printed books and drawings as being his private property, and all journals[2], accounts, correspondence,

[1] More than a month elapsed before he could obtain an interview with Lord Duncannon, and then only through his having been introduced at a breakfast party to one of the Commissioners, Mr. Milne, who arranged the meeting for him. He found his lordship, he writes, very communicative; he told him of the difficulties and obstacles, but that they were not insurmountable, that he would with the greatest pleasure further all Lord John's views to the utmost of his power, and that all he wished was, that there should be no more ground taken into the Gardens, and that the Civil List should not be further burthened.

[2] The journals, &c., were for the most part, I believe, transferred to the Commissioners on the death of Mr. Aiton, who had retained his official residence, and they are now at Kew. The collection of drawings, made under Mr. Aiton's supervision, was subsequently presented to the Royal Gardens by Mr. Atwell Smith, a relative of Mr. Aiton. There had been in the office a considerable herbarium of garden plants and of others made by collectors sent from the

and other documents as not being the property of the Commissioners. On April 1, 1841, my father received his commission, the acceptance of which was regarded by his friends as a very insecure foundation on which to build the object of his ambition, a Botanic Garden worthy of the nation. But he was confident of the support of the scientific public in whatever he should undertake, and, I suspect, of that of more than one of the Commissioners of Woods and Forests.

The next step was to find a residence within a reasonable distance from the Gardens. There was none to be had within two-thirds of a mile, where, in the adjoining parish of Mortlake, there stood a commodious three-storied many-roomed building, of which he took a lease. It was pleasantly situated on 7½ acres of ground with some fine trees that stretched down to the Thames, had a walled garden, orchard, stables, and coach-house, and was in good repair. It bore the name of Brick-stables, for which its owner, the possessor of large property in the vicinage, substituted that of West Park [1].

The translation from Glasgow to West Park occupied my father for three months, during which he was heavily and painfully handicapped by the absence of my mother, who was nursing a dying daughter in Jersey, and the illness of his father, who was nearer ninety than eighty years of age, and had lived with him for ten years. His only surviving son was serving in the Antarctic Expedition under Captain (afterwards Admiral Sir James Clarke) Ross. There being no railroad available in those days, he hired a smack for the conveyance by sea of his furniture, household goods and gods, herbarium and library, from Glasgow to London, where they were put into lighters and landed on the banks of the Thames at West Park itself. Previous to this he had lightened his library by the sale of 1,000 volumes, chiefly of classics, Delphine, Aldine, and Elzevir editions, collected in the middle of the previous

Gardens to Australia, the Cape, &c., but these had already been sent to the British Museum.

[1] West Park has disappeared. Its former site is occupied by the sewage works of Kew and Richmond.

century by his godfather, Mr. Jackson of Canterbury. The cost of the move was about £300, his first year's salary.

Early in July he was settled at West Park, where the drawing-room, ante-drawing-room, and study, were shelved from floor to ceiling and filled with books, and five rooms were occupied with the herbarium.

Nothing was allowed him for the conveyance and fittings necessary for these indispensable working materials[1], which he kept up mainly at his own cost, for the use of the establishment, for twenty-four years.

On entering upon his duties under the Commissioners of Woods and Forests the new Director was cordially welcomed, and to his surprise and gratification found that he had a free hand, and promise of favourable consideration in projecting improvements in the Botanic Gardens. His plan of operations is tersely and best given in his first Report presented to Parliament on the condition of the Gardens, which begins with, 'Having no instructions for my guidance I determined to follow the suggestions of Dr. Lindley's Report.' Meanwhile Lord Lincoln (afterwards fifth Duke of Newcastle) had succeeded Lord Duncannon, and in him, Mr. Milne, the Honourable C. Gore, and Mr. Philipps, secretary to the Board, he found gentlemen as interested as himself in the development of the establishment, who made frequent visits, going into every detail of garden works, and giving much of that 'efficient assistance in scientific management and adaptation to useful purpose' which their former Chief Commissioner had declared the Board to be incapable of affording.

To give a clear account of the additions made and improvements carried out in the establishment of Kew, it will be convenient to consider them as far as possible under the four heads of Botanic Gardens proper, Pleasure Ground or Arboretum, Museum, Herbarium and Library.

Botanic Gardens proper. The first recommendation of the new Director was that these should be open to visitors on

[1] In Dr. Lindley's Report, the necessity of a herbarium and library for the performance of the Garden duties was indicated.

week-day afternoons throughout the year; of which privilege upwards of 9,000 persons availed themselves during the remaining nine months of the year[1]. The next, in 1842, was that the permission of Her Majesty should be asked to add a few acres of the Pleasure Ground to the old Arboretum for the purpose of opening a new entrance[2] to the Gardens from Kew Green. This was graciously granted, as were the far larger areas from time to time asked for, of which the next (in 1843) was for forty-eight acres, to afford sites for a new Pinetum, and for the erection of a Palm House far exceeding in dimensions any previously constructed.

In 1846 the Royal Kitchen Gardens, which had remained under Mr. Aiton's management, were annexed to the Botanic Gardens. They occupied an area of about fifteen acres skirting the Richmond Road. A good-sized building used as a storehouse for fruit stood on this site, together with a large vinery and several forcing-houses, frames, melon and other pits. The vinery and some of the latter were in such disrepair as to be condemned, others were retained; but a considerable area being provided with excellent garden soil was devoted to the formation of a new collection of hardy herbaceous plants arranged according to the Natural System. This, according to a printed catalogue drawn up by Mr. Niven, foreman of the department, in 1853, contained about 5,500 species, a number no doubt swollen by the admission of half-hardy plants, varieties, and synonyms. The first hardy herbaceous collection in the Royal Gardens was formed in 1760, near the Temple of the Sun. It was an acre in extent, contained 2,712 species, and was called the Physic Garden. According to the first edition of Aiton's *Hortus Kewensis* (1789), there were about 2,824 hardy herbaceous plants cultivated in the Royal Botanic Gardens, and in the second edition (1810–1813) 3,946 species.

[1] In the last year of his Directorship (1865), Sundays having in the interim been included with the open days, 73,307 persons were admitted. In 1883 the numbers had risen to 1,240,489.

[2] The noble gates erected on this spot in 1845 are from designs by Decimus Burton, F.R.S.

Of the plant-houses existing in 1841 about ten were of considerable size, and of these two only, an Orangery and the architectural Conservatory near the entrance gates, are still (1902) standing. The Orangery was built by Sir W. Chambers in 1761 for wintering orange trees. It is the first plant-house of any importance erected in the grounds, and had been latterly used as a conservatory for the reception of such trees as had overgrown the height of the New Holland House. It was so occupied till the completion of the Temperate House in the Arboretum in 1863, when its contents were transferred to the latter, and replaced by Museum objects, as will be more fully described under Museums. It was almost the only remaining house heated by flues under the floor, the dry air from which was very unfavourable to the plants, as was the want of light.

The other permanent building was the architectural Conservatory near the entrance gates. It is one of a pair erected in 1836, the other is at Buckingham Palace. It was heated on Perkin's system of innumerable coils of pipes, the size of ordinary gas-pipes, charged with steam [1] from twelve furnaces in the vaults. It was used for the same purposes as was the Orangery, and its contents were similarly disposed of at the same time, when these were replaced by tropical plants, chiefly small palms, tree-ferns, and aroids, not a few of which are still flourishing. With regard to the other old houses, some were destroyed and better provided, others were improved and added to, and the majority had the old system of heating by hot air from flues passing under the floor replaced by that of hot-water pipes. In the case of six of these houses the flues from their furnaces were conducted into one shaft, thus contributing much to the cleanliness of their surroundings. Two span-roofed houses were doubled, one for tropical, especially economic plants, the other for Australian and New Zealand ones.

The building of the great Palm House was commenced in 1844 from the designs of Decimus Burton, F.R.S., and the

[1] At a later period this system was replaced by one of hot-water pipes.

Director. It was completed in 1848, together with the campanile, which was intended to serve the purposes of a smoke-shaft and water-tower, and the ornamental terrace facing the water. The flues of six furnaces which heated the boiler of the hot-water apparatus were carried in a tunnel to the base of the campanile, where there was a furnace for the consumption of the smoke and for securing a powerful draught[1]. The tunnel also served for the conveyance of fuel from the coke yard by the Richmond Road to the furnaces, thus avoiding the necessity of carting over the lawns. The dimensions of this building are—length 362 feet, centre 262 feet by 100 wide, and 66½ feet in height; the wings are each 50 feet long and 30 feet in height. It is glazed with about 45,000 square feet of sheet glass. A gallery runs round the central portion at a height of 30 feet; there are 19,500 feet of hot-water pipes 4 inches in diameter. It is, except the Crystal Palace and subsequently erected Temperate House at Kew, I believe, the largest glass house in existence.

At this time the activity of the Commissioners of Woods and Forests was far-reaching, for it was in their contemplation to annex the Chelsea Botanic Gardens to Kew and to form a Medical Garden for the use of the colleges and schools of London. Referring to these schemes in letters to Mr. Dawson Turner, my father in 1843 writes in respect of the formation of a Medical Garden: 'It will be attended with many difficulties, but I shall encourage it, and have written a long memorial to the Board about it.' In 1845 he writes: 'I have to write to the "Woods" on an affair to be laid before the Queen respecting a Medical Garden adjoining the Botanical Garden.' And again in 1845: 'My Report on the Gardens is printed by the House of Commons, and my letter on the removing Chelsea Garden to Kew. Lord Lincoln thinks it will result

[1] This arrangement proved unsatisfactory and had to be abandoned. The flues from the furnaces are now led to two shafts in the centre of the wings of the house, the projecting mouths of which are masked by octagon lanterns. This effected both a great saving of fuel and an increase of heat, but it was not till a double coil of pipes was led round the gallery in the central compartment that the house was heated sufficiently for its purpose.

in that Garden being removed here, or in Government forming here a Medical Garden on a national scale.' Both schemes were abandoned.

In 1846 a new Orchid House was built, and a wing 60 feet long was added to the New Holland House. The Orchid House was especially required for the accommodation, besides the Kew collection, of two others, one the Woburn Orchids, which had been presented by Duke Francis to Her Majesty with the view of their being transferred to Kew; the other that of the Rev. F. Clowes, of Broughton Hall, Manchester, a very large one, eminently rich in Andean species, generously presented by its owner.

In 1850 the Tropical Aquarium, or Water-lily House, was built, at a cost of about £2,000, chiefly for the cultivation of the *Victoria Regia.* It stands near to the north end of the Palm House, from one of the boilers underneath which it was heated[1].

In 1855 a Succulent House, 200 feet long by 30 feet wide, was erected and filled from end to end with Cactuses, Crassulaceae, Aloes, Agaves, S. African Mesembryanthemums, and allied plants of dry climates. This house became very popular, but it is difficult to say whether the visitors were more interested, or instructed, or puzzled, by the strange and novel forms it contained. One Cactus which lived for several years was the wonder of the Kew collections. It weighed one ton. It belonged to the genus *Echinocactus,* of which the species are more or less globular, and in this case was 9½ feet in girth. My father reported that he paid a bill of fifty guineas as the cost of its transport from the mountains to the coast of Mexico in a wagon drawn by six oxen. Owing to a bruise received in transit it very slowly rotted away.

Reverting to the greatly extended Botanic Gardens, now nearly 70 acres in extent, the laying out of the new ground,

[1] This arrangement did not answer; the loss of heat in transmission was too great, and eventually the house had to be supplied with a separate boiler and furnace.

construction of main walks for the accommodation of large crowds of visitors, and of subsidiary ones leading to the various plant-houses, and the designing of extensive geometric flower-beds in character with the Palm House, the gates, and other great structures, demanded the genius of a professional landscape gardener, and Mr. Nesfield of Eton was selected for the purpose. The conditions to be met presented many difficulties, the views in all directions were dominated by conspicuous insurmountable objects, the imposing entrance gates, the Palace, the Orangery, the campanile, the Pagoda, the Temple of the Sun, a piece of water, and two large artificial mounds crowned with classical temples. The result has proved satisfactory, its main features called forth no adverse criticism, and remain to this day almost as they were planned upwards of half a century ago.

As these works were progressing, Her Majesty, accompanied by the Prince Consort, in 1843, paid her first visit to the Gardens. They were pleased to express their approval of all they saw, and were especially interested in a model of the Palm House, then about to be erected. They subsequently sent the Royal children on several occasions that my father might point out to them the more interesting plants in the houses. The Prince Consort paid several subsequent visits, and took a keen practical interest in all that was doing in the Pleasure Ground, which was separated from the Botanic Gardens by a light wire fence, and in the Deer Park beyond. He took the finer trees under his especial protection and forbad the cutting down of any without his sanction. The Deer Park he declared should never be built upon, and he approved of my father removing the wall that separated it from the Pleasure Grounds.

Now that I am in duty bound to introduce Royalty into a sketch of my father's life it would be disloyalty, as well as ingratitude, to pass over the life-long connexion of the late Duke of Cambridge and his family with Kew. The gardens of Cambridge Cottage abutted on the Royal Botanic Gardens, and for a great part of the year the family resided there, and daily walked in the grounds, or the Duchess drove in a light pony carriage, carefully keeping off the lawns and the edges of the walks. The Duke died before my father took up his residence at Kew itself, but the Duchess and the Princess Mary, afterwards the Duchess of Teck, constantly invited him to accompany them in their walks, and were not backward in giving him their opinion of his operations. The Princess would come and tap at his study window for him to come out and show her interesting plants in the houses and grounds, though he never (as has been reported in some biographies of Her Royal Highness) either taught her botany, or was ever asked to do so. Knowing as all do her charm of character, it is not surprising that my father, who was no courtier, greatly enjoyed such interviews with his Royal neighbours; and he profited by them too, for he had the opportunity of meeting at Cambridge Cottage men of the highest distinction, and introducing them to the wonders of Kew.

The only other allusions to Royalty which I find in my father's correspondence are of a visit to Osborne by command of the Queen in 1850, and the following extract from a letter to Mr. Turner, dated August, 1854: 'The Gardens are increasing amazingly in beauty, interest, and popularity. The Queen has been here three times in less than six weeks, and I was required by her to inquire if the Palace could not be put in repair for her and her children. The Duchess of Gloucester[1] commanded my attendance last week; the Queen Dowager came on Friday with a very large suite, and remained three hours in the Gardens and Museum. The Palm House, now that it is filled, is the admiration of everybody, and the view of the palms from the gallery is most striking. The Queen was enchanted with it. But the Museum is, if possible, more attractive still, and is crowded daily.'

In 1843 my father reverted to the plan followed during the palmy days of Kew, when under the patronage of Sir

[1] The Duchess of Gloucester and Princess Sophia were the last of the Royal family to reside at the Palace of Kew.

Joseph Banks, of sending collectors to distant countries for the purpose of transmitting plants and seeds to the Royal Gardens; and by way of lightening the demand on the Treasury he on several occasions, with the Commissioners' approval, invited the Duke of Northumberland and the Earl of Derby to contribute to such expeditions and share the produce. At the same time, through his influence at the Admiralty, he obtained the privilege of having all packages addressed to Kew coming by the Royal Mail West India steam packet sent freight free. By these means Mr. Purdie was sent to New Grenada, and Burke and Geyer to California and Oregon, with the most satisfactory results to all parties; and by similar arrangements with the Treasury, Foreign, Indian, and Colonial Offices there were subsequently sent Oldham and Wilfred to Japan, Formosa, and Corea, Mann to the Cameroons, Gaboon River, and Fernando Po, Baikie and Barter to the Niger, Kirk to the Zambesi with Livingstone, Meller to East Africa and Madagascar, myself to the Himalaya, Bourgeau to Canada, Lyall to British Columbia, Edmonstone, followed by Seemann, to Western and Arctic America in H.M.S. *Herald*, and the latter to the Fiji Islands with Col. Smythe's mission, Macgillivray to Torres Straits in H.M.S. *Rattlesnake*, Milne to the Pacific in H.M.S. *Herald*, Spruce to Ecuador for Cinchona seeds, and Hewett Watson to the Azores. The practice was definitely abandoned when the great nurserymen took it up, and liberally shared their proceeds with Kew in exchange for its Director's services in indicating countries worth exploring, giving the collectors letters of recommendation to his correspondents abroad, naming and publishing their novelties and rarities, &c.

In 1844 my father was instructed to prepare a Guide-book to the Gardens for sale at the entrance, and to make an annual Report on the progress and condition of the Gardens, to be laid before Parliament. The first edition of the Guide-book contains fifty-six pages and sixty-one woodcuts of objects exhibited. It was entitled 'Kew Gardens, or a Popular Guide to the Royal Botanic Gardens of Kew,' by Sir W. J. Hooker,

Director. After bringing out twenty-one successive editions he transferred the duty to Prof. Oliver, the keeper of the Library and Herbarium, who in 1863 included the Arboretum in the Guide book.

In 1850 the Board of 'Woods and Forests' was divided into two, that of 'Works and Public Buildings,' of which latter the Chief Commissioner became a Minister of the Crown; and that of 'Woods, Forests, and Land Revenues,' which became practically a department of the Treasury. The Botanic Gardens and Pleasure Grounds were transferred to the first of these as revenue-expending establishments; the Deer Park to the second as revenue-yielding. The Palace and its domains, together with the Swiss Cottage in the Pleasure Ground and the ground around it, remained under the control of the Queen's Household[1].

On this fission of the office its chief, Lord Seymour (afterwards twelfth Duke of Somerset), became Chief Commissioner of Works and Public Buildings, of which Mr. Philipps retained the secretaryship. Mr. Milne, on the other hand, remained a Commissioner of Woods and Forests, &c., to my father's great regret, for this gentleman had from the first taken a special interest in the development of Kew Gardens, and had in many ways smoothed the Director's path. Happily in Lord Seymour and Mr. Philipps my father had excellent friends, and I have heard him say that Lord Seymour was on the whole, as a man of business and intelligence, the most efficient chief he had served under. I must add that as First Lord of the Admiralty the Duke continued to show his zeal for the establishment he had ruled over.

The following passages from my father's correspondence are interesting as referring to effects retained by Mr. Aiton (who died in 1849), on resigning the Directorship of Kew:—

'Kew, August 13, 1851. I have just received a catalogue of the *two* brothers Aiton's combined sale, which is to take

[1] Her late Majesty latterly transferred these also to the Department of Works and Public Buildings, when they came under the control of the Director of Kew.

place to-day and three following days at Kensington, and I am happy to tell you that the botanical portion (600 volumes), herbarium, botanical drawings, &c., are excluded; from which I can only infer that Lord Seymour has so far accepted my recommendation as either to have purchased them by private contract or to have arranged for their being kept back with a view to their being valued. Otherwise they were certainly to have been sold to-day, the books being all catalogued and numbered for the purpose by the auctioneer. It is odd that Lord Seymour should not have written further to me; but with best desire to do everything he can for the advantage of the Garden, he seems to feel that he is put into the office to keep a jealous watch over all the officers, and to take care there is no jobbing.'

'Sept. 2. In spite of Lord Seymour's refusing to purchase the books, &c., of the Aiton representatives, Mr. Atwell Smith has sent to me for the Garden Zoffany's very fine portrait of old William Aiton, and sundry MS. books which ought never to have been removed from the Garden. Lord Seymour little knows he has to thank me and the attentions I was able to pay to Mr. Atwell Smith by attending the two Aitons' funerals, &c., for these objects having come here at all. The picture is suspended in the Museum, and I am getting sixty of my large folio drawings (made for my Glasgow lectures) framed for the walls in the rooms now being added to the Museum.'

Referring to the above effects of the brothers Aiton, the books were sold, but the collection of drawings, which is of great value, was retained, and subsequently presented to Kew by Mr. Atwell Smith.

In 1853 a house in Kew, in possession of the Queen, having become vacant through the death of its tenant (Sir George Quentin, Riding-master to the family of George III), Her Majesty was pleased to place it at the disposal of the Commissioners of Works, to be in future the residence of the Director of the Botanic Gardens, in which it was situated. This was to my father a very great boon. He was in his sixty-ninth year, and burthened with the duty of creating a National Arboretum in the Pleasure Grounds, nearly two miles distant from West Park, and demanding unremitting scientific supervision. Nor must it be forgotten that his herbarium was outgrowing his accommodation for it, and that his expenses all along far exceeded his official salary[1]. The house was a good one, facing the Green, with its back in the Gardens, but it would not accommodate his library and herbarium, which, together with his study and artist's room, occupied thirteen apartments in West Park.

Fortunately a large house[2] closely adjacent to the Botanic Gardens, which had formerly been occupied by the King of Hanover, afforded abundant space for the herbarium and library, of which last he kept in his study such works as were in frequent use. A history of the Herbarium and Library at Kew during my father's lifetime will be found further on.

Returning to the operations in the Botanic Gardens, in about 1855 instructions were given to the Director (to his great discomfiture) to decorate the lawns and borders of the paths over a considerable area of the Botanic Gardens with 'carpet-beds' of flowers. These he regarded as out of place in a garden where objects of as great beauty, and far greater interest both popular and scientific, abounded. He further regretted the great expenditure on propagating-pits, frames, soil, and labour, on a show of but a few weeks' annual duration, whilst some scientific branches of the establishment

[1] On taking up his residence at Kew he was allowed to retain with his salary (by this time £500) the allowance for rent (£200) which he had at West Park; in 1855, after several appeals for aid in conducting his enormously increased duties, I was appointed Assistant Director of the Royal Gardens, Kew.

[2] This house, with the grounds around it, had belonged to a Mr. Hunter, from whom it was purchased in 1818 by King George III, at the instigation of Sir Joseph Banks, to provide for a Herbarium and Library to be attached to the Royal Botanic Gardens. The only objective evidence of its original destination was that one of the rooms was shelved for books. In 1823 George IV sold the house and grounds to the nation: in 1830 William IV granted it to the Duchess of Cumberland for her life. On the accession of the Duke of Cumberland to the throne of Hanover, it became known as 'The King of Hanover's house.' It is now entitled 'The Herbarium of the Royal Gardens.'

were being starved, and a structure of the dimensions at least of the Palm House, to rescue the magnificent collection of colonial trees, &c., from destruction or deformity, was urgently needed. The object of the proposed decorations seemed to be to rival the London parks, where such an attraction was eminently suitable and admirably carried out. In the end he came to an arrangement with his chief (Sir Benjamin Hall, I think), that a sum of money should be added to the estimates and appropriated to this decorative work, and that he be supplied with a skilled foreman to carry it out. The system was continued for several years, and was thereafter gradually suppressed.

The years 1860 to 1862 were notable for the successful efforts in introducing the Peruvian barks into India and our tropical colonies. Mr. (now Sir Clements) Markham had induced the Indian Government to undertake this measure, which had been urged upon it by Sir Joseph Banks more than half a century before, and by various botanists since. Mr. Markham himself went to Peru, and brought to England living plants, which, after a short nursing at Kew, he took on to India and established in the Nilghiri Hills. Meanwhile my father, to whom the Indian Government applied for advice, not trusting wholly to the risky transport of living plants, urged that collectors should be sent to Ecuador and Bolivia for seeds of the different species, recommending at the same time the employment in Ecuador of Mr. R. Spruce, an able botanist and collector, who happened at the time to be in that country.

In the Report on the progress and condition of the Royal Gardens during the year 1861 it is stated that:—'The means adopted for introducing Cinchonas (trees yielding quinine) into the East Indies and our tropical colonies rank first in point of interest and importance of the works of the past year. In my Report for 1860 I mentioned the erection, at the desire of the Secretary of State for India in Council, of a forcing-house, especially for the cultivation of the Cinchonas, with the view of establishing plantations of them in

India. The operations of the several parties organized to proceed into the Andes and procure young plants and seeds have been described in detailed reports laid before the Secretary of State for India by Clements R. Markham, Esq. Upon the Royal Gardens devolved the duties of receiving and transmitting the seeds and plants to India, of raising a large crop of seedlings, of nursing the young stock, lest those sent on should perish or the seeds lose their vitality, and of recommending competent gardeners to take charge of the living plants from their native forests to the hill country of India, and to have the care of the new plantations there. Further, with the sanction of the Indian and Colonial Governments, it was arranged that our West Indian colonies and Ceylon should be supplied with a portion of the seeds.'

In the Report for 1862 the number of plants established in the Nilghiris is 117,706; in the Sikkim Himalaya 2,000 [1], in Ceylon about 3,000. In the Report for 1863 the number of plants in the Nilghiris is stated to be 259,356, and in the Himalaya 8,000, where applications have been made to the superintendent of the plantation from private individuals for 1,500,000 plants; in Ceylon 22,050.

In 1864 and subsequently great efforts were made to introduce the Ipecacuanha plant into India from Brazil, but with little success. The plant was impatient of removal from its native forest and of transportation, and wâs further one of extraordinary slow growth. Such specimens as arrived in India in a living state made no progress, and the attempt had to be abandoned.

In 1861 a reading-room and some books and horticultural journals were provided for the gardeners, and Professor Oliver, keeper of the Library and Herbarium, volunteered a course of elementary lectures on botany. It was not till some ten

[1] Manufactories of quinine have been established in the Sikkim Himalaya and the Nilghiri Hills. The most signal proof of the success of the experiment is, that a dose of five grains of quinine in a paper bearing a Government stamp may be bought at any post office in Bengal for half a farthing. This supply is from the Sikkim manufactory.

years had elapsed that a system of paid lectures was organized, which have proved a great boon and success.

In 1865, the last year of my father's life, he received from the Lords of the Admiralty the gratifying intelligence, that his long-sustained exertions in sending such plants to the sterile Island of Ascension as would most effectively and speedily clothe its naked soil, and thus conserve a water supply, had been crowned with success. It was in 1843, after the return of Sir James Ross's Antarctic Expedition, which had touched at the island on its homeward voyage, that the idea of planting that island extensively with such trees, herbs, and shrubs as were best suited to its soil and climate originated. Ascension being a naval station, the Admiralty favoured the idea, and Kew was applied to for aid in giving effect to it; which it did by sending out seeds and cases of living plants year after year, and a succession of young gardeners to plant and sow. According to Captain Barnard's Report to the Admiralty, 'the island in 1865 possessed thickets of upwards of forty kinds of trees, besides numerous shrubs and fruit trees, of which, however, only the Guava ripens. These afford timber for fencing cattle-yards.' In 1843 there was but one tree on the island and no shrubs, and there were not enough vegetables produced to supply the Commandant's table. The Report goes on to say: 'Through the spread of vegetation the water supply is excellent, and the garrison and the ships visiting the island are supplied with abundance of vegetables of various kinds.'

The *Arboretum*, formerly the *Royal Pleasure Grounds of Kew*. In 1845 Mr. Aiton[1] was relieved of the charge of that portion of the Pleasure Grounds (about 178 acres) then in occupation of the King of Hanover as a game preserve, which had not been as yet added to the Botanic Gardens, together with the Deer Park (350 acres), and my father was asked to

[1] Mr. Aiton, on retiring with a pension of £1,000 per annum, had begged to be allowed to retain the Directorship of some portion of his realm, on the very natural plea that his services under the Royal Family might, if possible, be life-long. He died in Kensington, October 9, 1849, in his eighty-fourth year.

include these in his Directorate. This he agreed to do, though no hint of an increase of salary accompanied the request, and though the duties involved were neither botanical nor horticultural, but rather agricultural. He had no doubt two good reasons for this compliance, one in having an eye to the remainder of the Pleasure Ground as the site for an Arboretum worthy of the nation; the other, that to have allowed these to be placed under any other authority might have led to complications.

Thus the Director's rule was extended in four years from a Botanic Garden of eighteen acres and a few hundred yards in length, to an area of nearly 650 acres, extending from Kew Green to the Thames at Richmond, two miles distant. Some idea may be formed of the labour which this acceptance of extra duty entailed from the following extract of a letter dated March, 1846, and addressed to Mr. Turner: he says, 'For myself the Gardens have never made such demands on my time as at the present season, when the most extensive operations are being carried on in the Pleasure Ground, as well as in the Botanic Gardens. In each place our usual complement of men is much more than doubled. In the former, owing to the severe illness of the foreman, I have to superintend everything, and there is literally not a man in whom I can put confidence about the place. I have lately detected very gross abuses, which there is every reason to believe have been practised for a long time under the régime of my predecessor.'

With regard to the Deer Park, often then called Richmond Old Park, except for planting some clumps of trees and shrubs, it suffered no change so long as my father had charge of it. It was let to a grazier and yielded large crops of hay. The Observatory[1], which stands towards the centre of it, was

[1] The Observatory was erected in 1798 by George III for the purpose of observing the transit of Venus, and for the instruction of the younger members of the Royal family in astronomy. For many years it was devoted to scientific purposes, under the direction of accomplished astronomers; and served for regulating the clocks in the Horse Guards, St. James's Palace, and elsewhere in London. In 1840 its contents were dispersed, and the principal instruments sent to King's

then unoccupied, but shortly afterwards it was placed at the disposal of the Meteorological Committee of the Royal Society.

In the Pleasure Grounds the only improvement at first effected was the removal of a wall about three-quarters of a mile long which separated them from the Deer Park, and replacing it by a ha-ha, thus opening the latter to view, together with the distant views of Isleworth Church, Sion House, and the woods on the opposite side of the Thames.

Then followed the formation of avenues in the grounds, upon which point Mr. Nesfield was consulted, the construction of paths, and the establishment of a nursery for the purpose of rearing accessions to the Arboretum and projected Temperate House or Winter Garden, and of propagating duplicates for distribution and exchange.

In the formation of the Arboretum, which occupied about three years, some of the principal nurseries of the United Kingdom and the Continent were laid under contribution for specimens often of great market value, and it is impossible to exaggerate the liberal spirit with which the owners of these responded to my father's call. All seemed to recognize the national character of the work, and that, as a means of enabling them to verify scientifically the nomenclature of their stock in trade, its services would be invaluable. The two most conspicuous and beautiful features in the Arboretum formed at this time were the Rhododendron walk and the Azalea beds. The number of species and marked varieties in the Arboretum was 3,500, grouped under their natural orders and genera.

College, London. It is well that the Botanic Gardens did not share the same fate.

I may here remind my readers that Kew has claims for the worship of astronomers, as well as of botanists, for that in a house which stood opposite the Palace, and which was taken down in 1803, resided the celebrated astronomer Samuel Molyneux, F.R.S., secretary to George II when Prince of Wales. It was with a telescope constructed by Mr. Molyneux and placed on the lawn near his house, that Dr. Bradley made in 1725 the first observations that led to his two great discoveries of the aberration of light, and the nutation of the earth's axis. To perpetuate the memory of so important a station, His Majesty King William IV had a sundial with a suitable inscription placed on the spot where the telescope had stood. It is no doubt to the fact of His Majesty's having had the education of a naval officer that this rare tribute to a scientific man and his discoveries was due.

Another nursery was established in 1855 in the private grounds near the Palace, for the very different purpose of supplying the London and other parks, the property of the Crown, with trees and shrubs. Up to the year 1865 more than 25,000 trees and shrubs had been supplied from these two nurseries, many to Battersea, Hyde, Victoria, and Richmond Parks; others to plant open spaces in the private grounds of the Queen around the Swiss Cottage for the encouragement of wild birds, and to form a belt a quarter of a mile long by the banks of the Thames to screen the Arboretum from Brentford. Many wagon-loads were sent to Aldershot, Deptford, and other yards.

In 1857 a lake four and a half acres in area was formed in a marshy depression that had communicated with the Thames opposite Sion House, and which had been enlarged and deepened by the removal of many cart-loads of gravel for the formation of paths, and for the terrace on which to place the Temperate House. The lake was finished in 1861, its banks planted, and a communication with the Thames re-established by a tunnel and sluices. It has been enlarged considerably in later years.

By far the greatest undertaking carried out in the Arboretum was the construction of the main body and octagons of the Temperate House, or Winter Garden, as it was at first proposed to call it. From 1856 onwards my father had annually urged on the Government the necessity for such a building, and I cannot do better than reproduce the words of his Report for 1857, 'On the condition of the Royal Gardens,' as showing cause for its construction not being delayed:—'All the plant-houses are progressing favourably, with one exception, to which I have already alluded, as a source of deep concern. Unless we have, at once, a structure suited to the reception of our large trees and shrubs which will not bear frost, especially that once celebrated collection of pines, Araucarias, Proteas, &c., they will soon be past recovery. Already they have suffered extremely for want of space; many have perished, many are deformed and crippled, being shorn every now and then of their graceful and stately heads in order to bring them under the shelter of a dark roof, that of the "Orangery," only twenty-three feet high, or in a hovel of a building long ago condemned as discreditable to the Gardens. The crying need of a new Conservatory has long been admitted. One very old and decayed greenhouse, which had been tenanted by a portion of the very plants in question, was pulled down four years ago with the understanding that it should be replaced by a better building; and numerous desirable works have been postponed that the money destined for them might be applied to erecting a structure commensurate to our wants. I do not know that I can express my views on this subject in stronger words than I used last year, and which I beg to repeat. In my Report for 1856 I said that I must speak almost in the past tense of those superb Mexican, Australian, and Norfolk Island Araucarias, conifers, &c., which were once the pride of Kew Gardens, but that while some had suffered past recovery, others might still be restored by affording them needful space, light, and temperature. I added, as a further proof of the evident necessity of the house in question, that during the whole sixteen (now seventeen) years of my Directorship, not any addition had been made to the accommodation for these kinds of plants; it had indeed, as above shown, sustained a diminution. The Gardens cannot be deemed complete till the trees and shrubs of temperate climates are as well cared for as the tropical plants, for whose reception our noble Palm House was erected thirteen years ago. Then, and not till then, will the national establishment be perfect. A botanical garden is not valuable, as was once thought, for the number, mainly, of the species which it includes, but for their usefulness and beauty; they should be a *selection* rather than a *collection*. The Conservatory in question would certainly cost a large sum of money, but not *nearly* so much as did the Palm House, which involved several items not requisite in a structure for hardier trees and plants. The price of glass, too, has fallen materially since 1844.'

In 1859 the design for such a conservatory, by Mr. Decimus Burton and my father, was approved. It consisted of a central building 212 feet long by 137 feet broad and 60 feet high, with a gallery running round it at a height of 30 feet; two wings, each 112 feet 6 inches long by 62 feet 6 inches broad, and 37 feet 9 inches high; and two octagons interposed between the centre and the wings, each 50 feet broad and long by 25 feet high. In 1860 tenders were accepted for the construction of the centre and octagons only, and the work was at once proceeded with. In 1861 the octagons were completed and filled with plants in tubs and pots from the Orangery, the old Conservatory, and the architectural house near the gates, all in the Botanic Gardens; and in 1862 the centre was completed and its floor provided with beds, in which the larger specimens from the above-named houses were planted. Unfortunately no representations availed to induce the Government to complete the building by the erection of the wings, the unoccupied naked gravel platforms for which were an eyesore to the Director for the remaining few years of his life[1].

The history of one more conspicuous feature in the Arboretum remains to be told. I allude to the stately flagstaff of Douglas fir, or rightly speaking flagstaffs, for there were two of them, though for obvious reasons the acquirement and fate of the first were not officially made public. In March, 1859, the Director received a letter from Mr. Edward Stamp, a gentleman engaged in the timber trade of British Columbia, offering to present to the Gardens a flag-staff of the Douglas fir, over 100 feet in height and not exceeding sixteen inches in diameter at the base. The offer was accepted by the Commissioners, and an excellent site fixed for it, on a mound in the Arboretum, not far from the Richmond Road, where it would be visible from the latter as well as from a great extent of both the Botanic Garden and the Arboretum. The spar, fully rigged for erection, was dispatched to Kew from the London Docks, floating, towed by a tiny steamer, but was wrecked

[1] They have since been erected.

en route, having been cut in two by another boat. Nothing daunted, Mr. Stamp had it back and the two pieces 'scarped,' thus reducing its length by a few feet. It was then again committed to the Thames, and safely landed on the river bank opposite Sion House, from whence it was transferred to the mound, in the top of which a suitable well had been sunk for the reception of the butt. The question of the proper person to entrust with the hoisting had been discussed with the Commissioners, when my father's suggestion of application being made to the Lords of the Admiralty for an experienced man from one of the ship-building yards was overruled in favour of the Clerk of the Works of the Office. The decision was unfortunate. The method adopted was to erect a derrick over the well, sling the spar, securely guyed, raise it horizontally to the required height, and then by depressing the end bring it to a vertical position and lower it. The occasion was a memorable one; a large party, including Royalty, was assembled to witness the operation, which resulted in a puff of wind striking the spar when in mid air, and bringing it and the derrick to the ground, where the spar lay broken into three pieces. I was present on the occasion and shared in full my father's vexation and bitter disappointment. The hardest task remained, the communicating the Commissioners' regret, together with his own, to the generous donor of the spar, who promptly answered with the offer of sending a longer one on his return to British Columbia! This he did, and in 1861 a spar about 250 years old, when felled 159 feet long, and 20 inches in diameter at the butt, rigged and ready for erection, was landed at the same spot as the former one had been. On this occasion the Director's suggestion was followed, the First Lord of the Admiralty (the Duke of Somerset, who as Lord Seymour had been first Commissioner of Works, &c.) was applied to, and a gang of riggers was supplied from Deptford Dockyard. These carried the spar to the mound, there laid it down with its butt in position in the base of the well (which was reached through a cleft in the mound), and then tilted it up to the perpendicular.

The show concluded by one of the riggers offering to stand on the truck of the spar for the gratification of H.R.H. the Duchess of Cambridge, who was present on the occasion, but who declined the offer. It should be recorded that all expenses attending the transport of both spars from their native forest to Kew, together with their dressing and rigging, were borne by Mr. Stamp, and that this second spar is believed to be the finest in Europe.

Museums. Referring to the storehouse for fruit in the old kitchen garden of Kew, alluded to at p. lvii as left standing in 1846 when that piece of ground was added to the Botanic Gardens, it appeared to my father that it might be converted into a Museum of Economic Products of the vegetable kingdom, raw and manufactured, and for the exhibition of large fruits and other objects of varied interests, nowhere displayed to view. Of such objects he had a large collection, formed chiefly for the use of his class in Glasgow, and others were scattered about the offices of the Gardens, some of them being the property of Mr. Smith, the Curator. Procuring a few trestles and planks, he formed of them a long table in the central room of the building, arranged all these articles on it, ticketed them, and invited the Commissioners to come and see them. This they did (I happened to be present on the occasion), and listened to his eloquent discourse upon them, during which he showed how such a collection of vegetable products might, besides interesting and instructing the public, prove of great service to the scientific botanist, the physician, the merchant, the manufacturer, the chemist and druggist, the dyer, and to artisans of every description. All these might find in such a collection the raw material (and to a certain extent the manufactured article) employed in their several professions, trades, or arts, correctly named, together with their native country and some account of their history.

The suggestion was adopted by the Commissioners, and, being approved by the Treasury, the room was fitted with glazed cases filled with objects and opened to the public in 1848, as the first Museum of Economic Botany ever formed.

The Museum was no sooner sanctioned than my father began to cater for objects wherewith to fill it, by application to merchants and manufacturers and by interesting his correspondents all over the world, many of whose replies were published in the 'Journals of Botany' which he conducted. His enthusiasm was catching. The Secretary of State for Foreign Affairs (Lord Aberdeen), on hearing of it, caused circulars to be sent to our Ministers and Consuls in foreign countries, desiring them to transmit specimens intended for Kew; the First Lord of the Admiralty (Lord Auckland) requested him to draw up instructions[1] for collecting for officers in their service, and the Minister of the Colonies displayed the same interest.

Consequently contributions poured in in embarrassing quantities, especially on the close of the Great Exhibition of 1851, when Messrs. Lawson and Co. of Edinburgh presented their magnificent exhibit of the agricultural products of Scotland. Increased accommodation was hence necessary and was found in two wings of the building, which had been used (one or both) as dwellings of gardeners; these provided four additional rooms and a staircase to a gallery which was constructed in the main room, and which was lighted from the roof. These completed the Museum building, which presented within 6,000 square feet of glazed wall-cases, and eight glazed table-cases, most of them with glazed drawers underneath. To add to the interest of the exhibits, framed coloured drawings of economic plants, palms, &c., lent by the Director, were hung to the gallery rails all round. A Guide to the Museum was drawn up by him in 1855, for sale at the Garden gates; it contained a plan of the interior, eighty pages of descriptive matter, twenty-six woodcuts, and notices of 560 of the objects most worthy of the visitor's inspection[2]. In 1857, when the

[1] The Admiralty Manual of Scientific Enquiry was the result, the botanical part of which was supplied by W. J. Hooker, assisted by D. Hanbury, Esq.

[2] In 1857, on the opening of the second Museum, he published a second edition of the Guide-book, in which 612 of the most interesting objects were described. In 1861 Professor Oliver brought out a third edition with notices of 1,000 exhibits.

larger Museum now to be described was opened, this one was numbered II in the Guide-book. Before proceeding to describe the second and third Museums erected by my father, it is gratifying to relate that within six years of the first being opened eight others, professedly on the lines of that at Kew, were established; they were in Edinburgh, in the India House (London), in Guiana, Jamaica, Melbourne, Calcutta, Madras, and in the Jardin des Plantes, Paris.

In the summer of 1855 the Director was invited by the Imperial Commissioners of the French International Exhibition of that year to take part in its functions, which resulted in his obtaining almost the entire collection of vegetable products there brought together. In aid of this he procured a grant of £400 from the Treasury, which the President of the Board of Trade, unasked, supplemented with a like sum. Thus provided, and with the ready assistance of the officers of the Board of Trade, and of the Science and Art Department, and enriched by donations of many exhibitors, he secured and transmitted to Kew forty-eight large cases of museum articles. This accumulation, and the facts that the Museum of 1848 was already overcrowded, and that great stores of specimens were being huddled away in the temples and sheds of the Gardens, led to the erection of a second and much larger building. This, which is Museum No. I of the Guide-books, was sanctioned by Parliament in 1854, was completed, fitted with 13,000 square feet of glazed cases, filled, and opened to the public in 1857. It is that now standing opposite the Palm House with the piece of water intervening. The expedience of following a classification of the contents of both Museums according to the Natural System necessitated the breaking up of the contents of the first, in which were retained all products of the Monocotyledonous and Cryptogamic divisions of the vegetable kingdom; the Dicotyledonous being transferred to the new building. In this laborious task the Director had the gratuitous aid of the Rev. Professor Henslow of Cambridge (Rector of Hitcham in Suffolk), who to his knowledge of botany and vegetable products, added singular skill in preparing and mounting the latter for exhibition[1]. His bust in marble (the gift of his sister) and that of the Director (the gift of Henry Christy, Esq.), both by Woolner, stand in the entrance hall of the new building, where are also, hung on the walls and projecting ends of the cases, a collection of nearly one hundred framed portraits of botanists[2], then the property of the Director. The third Museum (No. III of Guide-book), opened in 1863, originated in the timely conversion to this purpose of the Orangery, the oldest building in the establishment, 145 feet long in the interior. Timely it was in two senses; for the Orangery had hitherto been the main receptacle for such trees of Australia and New Zealand as had outgrown the old New Holland House, and in this very year the Temperate House was ready for their reception; also in 1862 my father had, thanks to the Secretary of State for the Colonies, the Duke of Newcastle (his first chief when Lord Lincoln), and to the Governors of several of the colonies themselves, acquired for Kew almost the whole of the vegetable products exhibited in the International Exhibition of that year, the East Indies being the chief exception. The colonies were, West Africa, the Cape of Good Hope, Natal, Mauritius, St. Helena, West Australia, New South Wales, Victoria, Tasmania, Queensland, Canada, New Brunswick, Vancouver's Island, British Columbia, Ceylon, Trinidad, the Bahamas, Dominica; together with these were exhibits from the Ionian Islands, Austria, Russia, and miscellaneous articles from other countries. Conspicuous amongst these acquisitions was the collection of colonial timbers, many as slabs of large

[1] Nor should the services on these occasions of the keeper for ten years of both Museums be forgotten—Mr. Alexander Smith (son of the Curator of the Botanical Gardens), who had acquired a remarkable knowledge of vegetable products. Owing to his health breaking down he was obliged to retire in 1858. He died at Kew in 1864. He was succeeded by Mr. John Jackson, who showed equal ability, and who in the course of his forty-two years of keepership became a leading authority on vegetable products. He retired on his well-earned pension in 1901.

[2] There are now (1902) about 190 portraits of botanists in this No. 1 Museum. There are also in the Herbarium 275 mounted in portfolios, and a few hanging on the walls, amongst which latter are excellent oil paintings of the late Mr. Bentham and of Professors Oliver and Baker.

size, selected from sound trees, partially polished and often of uncommon beauty. Almost all these timbers were named by men of scientific attainments and practical knowledge, and they were accompanied by reports containing a vast amount of serviceable information on their uses, qualities, &c.

Herbarium and Library. As stated at p. lvi, when the new Director of Kew took up his appointment, neither books nor a herbarium were provided for him; but he was well equipped with those of his own; nor was it till he was moved into a residence in the Royal Gardens, that he received any other substantial aid towards their upkeep and increase than house-rent, and latterly stationery and some cabinets. It is also told that the new residence not affording that accommodation for these which the Government had guaranteed, they were placed in a building adjacent to the Botanic Gardens. On this occasion it was arranged between the Commissioners and my father, that, on the condition of his herbarium and library being accessible to botanists [1], he should be provided with such a scientific herbarium curator as he had himself hitherto salaried [2].

Four years afterwards, the Royal Gardens came into possession, by gift, of the very extensive library and herbarium of G. Bentham, Esq., F.R.S., which was second to my father's alone in England in extent, methodical arrangement, and nomenclature, and which was placed in the same building. Its formation was begun in 1816, in France, where and in the Pyrenees Mr. Bentham collected diligently; but its great expansion by the inclusion of exotic plants dated from his introduction to my father in Glasgow in 1823, when the friendship between the two commenced which remained

[1] From the date of his taking up the Glasgow Professorship, his herbarium and library had been open to botanists, as was its owner's hospitable table to visitors from a distance.

[2] One of his curators, Dr. J. E. Planchon, subsequently attained to great eminence as Professor of Botany in Montpellier, where he carried out his researches in the vine disease caused by the ravages of the *Phylloxera*, which has cost France so many millions. He was the discoverer of the only effectual check to the propagation of this pest, by grafting *Vitis vinifera* on stocks of American species, which he proved to be almost immnne from the attacks of that insect.

undisturbed for forty-two years[1]. From that date the two botanists may be said to have hunted in couples for the aggrandizement of their libraries and collections, sharing their duplicates, Mr. Bentham giving my father the preference in all cases of purchase, &c. The one great difference between their aims was, that the former confined his herbarium to flowering plants, whilst my father's rapidly grew to be the richest in the world in both flowering and flowerless plants. The offer of this gift was prearranged with my father, who with his wonted disinterestedness put aside the obvious fact, that its acceptance would greatly diminish the value of his own herbarium and library, should the Government ever contemplate its purchase[2].

The principal additions to the Herbarium and Library made during the last ten years of the Director's life were:—

(1) The large collection made in North-West India, Kashmere, and Little Thibet by Dr. Thomson, and in the East Himalaya, the Khasia, Mount Silhet, and Chittagong by Dr. Thomson and myself.

(2) In 1858 seven wagon-loads of collections from the cellars of the India House in Leadenhall Street, where they had been accumulating for many years. They arrived at Kew in the chests in which they had been packed in India, many of them partly open and their contents destroyed by vermin and damp. Amongst the most valuable of these herbaria were those of Falconer in the North-West Himalaya (in the worst condition), of Griffith in Afghanistan, Assam, Bhotan, Burma, and the Malay Peninsula, and of Helfer in Tenasserim.

(3) In 1862 the herbarium of W. Borrer, F.R.S., long the Nestor of British botanists, and the life-long friend

[1] See Annals of Botany, vol. xii, p. 7.

[2] My father's herbarium had been offered to Government on several occasions, for a sum far below its value. After his death, it was (in 1866) purchased, with all such books, about 1,000 volumes (some of great rarity), as were not in Bentham's gift, together with a unique collection of botanical drawings, maps, MSS., portraits of botanists, and letters from his botanical correspondents from 1806–65, which amount to about 27,000.

and correspondent of my father, was presented by his widow[1].

(4) The Australian herbaria of Allan Cunningham, formed during that traveller's exploration of the interior of New South Wales and Queensland, made by himself in 1836 and 1838, and that made by his brother Richard Cunningham in 1835. Presented by R. Heward, Esq., F.L.S., of Kensington, in 1863.

(5) Mrs. Griffith's collection of British Algae. Presented by the Baroness Burdett Coutts in 1864.

(6) The specimens and original folio drawings, published and unpublished, upwards of 1,300 in number, illustrative of Dr. Boott's great work on the genus Carex. Presented by his widow in 1864.

(7) Dr. Lindley's Orchid herbarium, containing types of his 'Genera and Species of Orchideous Plants,' of his 'Folia Orchidacea,' and other works. Purchased in 1865.

(8) The immense herbaria of the traveller and naturalist Dr. Burchell, F.L.S., made in St. Helena, 1805–10, in South Africa (from the Cape to the Transvaal in 1811–5), and in Brazil in 1825–9. Estimated to contain 15,000 species, accurately ticketed for habitats and dates. Presented by his sister in 1864.

Turning now to my father's concluding botanical labours, the last of his efforts, the results of which have been far-reaching, was to address in 1863 a powerful appeal to H.M. Secretary of State for the Colonies, the Duke of Newcastle, K.G., in favour of H.M. Government undertaking to assist in the preparation and publication of a series of Floras of our colonial and Indian possessions. At the same time, for the information of the Secretary of State, he, in conference with Mr. Bentham, drew up and submitted the following estimate of the scope and cost of such a series of Floras,

[1] The Hookerian correspondence in the Herbarium at Kew contains 145 letters from Mr. Borrer, dated from 1823 only. I am indebted to Miss Borrer, of Brookhill, Cowfield, Horsham, for a series of 139 letters dating from 1803 to 1839, addressed by my grandfather, Mr. Dawson Turner, to her grandfather, in which there are frequent references to my father which have been of great service to me in compiling this sketch.

which is interesting as giving the views of the two best informed botanists in Europe as to the number of species of flowering plants and ferns natives of the several colonies, specimens of which were assumed to be available in herbaria for description at that time.

Estimated numbers of species to be described:—

Australian Colonies,	8,000	Hong Kong,	1,000
South Africa,	10,000	Mauritius & Seychelles,	1,000
British North America,	2,000	British Guiana,	2,000
West Indies,	2,000	Honduras,	1,500
New Zealand,	1,200	West Africa,	2,000
Ceylon,	3,000	British India,	12,000

Of these colonies the Flora of one only had, previous to the appeal to the Secretary of State, been completed on the plan proposed; that of Hong Kong, by Bentham in 1861. Three others were in progress, and have since been completed; the 'Flora of the British West Indies' by Grisebach, 1859–64; the 'Handbook of the Flora of New Zealand,' 1864–7, which includes all the known Cryptogams of the island up to date; and the 'Flora Australiensis' of Bentham, commenced in 1863 and concluded in seven volumes in 1878. One other Flora was in progress, and is not yet completed; the 'Flora Capensis' of Harvey and Sonder, of which three volumes were published between 1859 and 1865. All the above works were subsidized by the home or colonial Governments.

The number of volumes required was estimated to be forty-three; the author's remuneration to be £150 per volume, payable at date of publication. The price proposed was £1 per volume containing not fewer than 500 species. To insure the publisher against loss, 100 copies were to be taken by Government on the day of publication. The authors were to have no pecuniary interest in the sales of the volumes. The Floras were to be limited to flowering plants, ferns, and their allies, and to be written in English.

Of the botanical works published by my father during the twenty-four years of his Directorship of Kew, the more important were, the continuation of the 'Botanical Magazine,' volumes lxvii to xc, with 1,440 plates; the 'Icones Plantarum,' volumes iv to x, 700 plates; the 'Journal of Botany,' volumes iii and iv, with 28 plates; the 'London Journal of Botany,' 7 volumes, with 166 plates; the 'Journal of Botany and Kew Gardens Miscellany [1],' 9 volumes, with 109 plates. On ferns alone there were the 'Species Filicum,' 5 volumes, with 304 plates illustrative of 526 species; 'Filices Exoticae,' 100 plates; 'A Second Century of Ferns,' 100 plates [2]; the 'British Ferns and their Allies,' 66 plates; 'Garden Ferns,' 64 plates; and lastly, a commencement of a 'Synopsis Filicum.' To these must be added his Guide-books to the Royal Gardens and to their Museums, and his annual 'Reports,' to be laid before Parliament, on 'the progress and condition of the Royal Gardens.'

Altogether, inclusive of the 'Icones Filicum' (in association with Dr. Greville), my father published upwards of 1,200 plates of ferns, and descriptions of 2,500 species.

Alphonse de Candolle, in his warm tribute to my father's memory (Archives des Sciences de la Bibliothèque universelle de Genève, July, 1866), gives 4,094 as the number of plates of plants published by my father, exclusive of those in the 'Flora Londinensis' (about 220). This is far short of the total number, which I make to be nearer 8,000; of which about 1,800 were from drawings executed by himself. I need hardly add that but for the fidelity, artistic skill, and extraordinary rapidity of execution of Walter Fitch, who was my father's botanical limner for thirty years, this number could not have been approached.

With the commencement of a 'Synopsis Filicum,' which the completed 'Species Filicum' made a comparatively easy task, my father's labours terminated. His end was unex-

[1] This work was brought to a conclusion in 1857 with an almost pathetic farewell to the botanical helpers in his series of Journals, and to his botanical friends.

[2] The first Century of Ferns consisted of a re-issue of the plates and descriptions of the tenth volume of the Icones Plantarum (which volume was confined to ferns), on a larger sized paper, and coloured.

pected. On the Monday forenoon he spent two hours with me in inspecting Battersea Park, then in formation; here he left me and walked part of the way back to Kew, meeting by appointment the Queen of the Sandwich Islands and the Rev. Mr. Berkeley, with both of whom he spent the whole afternoon in the Gardens. On Tuesday morning his servant came to tell me that his master could not swallow. I followed immediately, and found him perfectly well except for this paralysis of the muscles of deglutition. I at once sent to London for the best advice, but to no purpose. I saw him no more, for sleeping on the floor by his bedside that night, under an open window, I was suddenly prostrated with rheumatic fever. Meanwhile he gradually sank, suffering no pain nor feeling the want of nourishment; and died from exhaustion, Saturday, August 12, in his eighty-first year [1]. He was buried in the churchyard of St. Anne's, Kew. A handsome tablet in the church with a central medallion profile by Woolner, and spandrels with groups of ferns in the corners, all in Wedgwood ware, record the dates of his birth, death, &c., with the motto, 'Thou, Lord, hast made me glad through Thy works.'

In person Sir William was over six feet high, erect, slim, muscular [2]; forehead broad and high, but receding, hair nearly black, complexion sanguine, eyes brown, nose aquiline—had been broken in a school fight; his mobile face, and especially mouth, was the despair of artists. Many chalk portraits of him were taken for friends by Sir Daniel Macnee [3], of which

[1] I have given these details because some of the published statements regarding the cause of his decease are erroneous.

[2] He was a vigorous pedestrian, covering 60 miles a day with ease. When taking the week's end rest at Helensburgh, during his summer course of lectures, he habitually on Sunday walked to Glasgow, 22 miles, to be in time for his 8 o'clock Monday morning class.

[3] Macnee was a youth of fourteen living in Glasgow when my father, who was one of his earliest patrons, went there. He made for the latter chalk portraits of Arnott, Bentham, Allan and Richard Cunningham, Douglas, T. Drummond, Greville, A. Gray, Harvey, Richardson, Torrey, Wallich, and Wight; all now hanging in the Museum of the Royal Gardens. Sir D. Macnee rose to be President of the Royal Scottish Academy of Arts in 1876. He died in 1882.

the best known to me is that which prefaces this article. Other portraits of him are two life-size in oil by Thomas Phillips, R.A., one in my possession, and the other in that of Sir Leonard Lyell, Bart., of Kinnordy; the half-length in oil by Gambardella, in the Linnean Society's meeting room; a small engraving in the series of portraits of members of the Athenæum Club[1]; one by Maguire in the Ipswich series of portraits of scientific men; and an etching in profile by Mrs. Dawson Turner, from a profile by Cotman, unpublished, but widely distributed. There is also the bust in marble by Woolner in the Kew Museum, an excellent likeness.

His general health was excellent, but he suffered from deafness, and sometimes serious trouble in one ear, brought on by an attack of scarlet fever in Glasgow, when also his throat was severely cauterized, an operation which left that organ very susceptible to cold. His habits were of the simplest; he was at work by eight a.m., and again till near midnight. Under medical advice he dined for the last twenty years in the middle of the day, and took a light supper at seven or eight. Afternoon teas were unknown in those days. It rarely happened that the midday dinner was not also the lunch of some expected or unexpected guest or guests. His absence from London society, and especially from meetings of the Royal, Linnean, and Antiquarian Societies, of all which he had been a member for fifty years, was greatly regretted. But these were held at night, at a distance of seven miles from his dwelling-house, and for the ten years of his West Park life an omnibus that passed quite half a mile off was the only public conveyance from Kew to the metropolis.

He was a Fellow of the Royal, Linnean, Antiquarian, and Royal Geographical Societies, LL.D. of Glasgow, D.C.L. of Oxford, a Correspondent of the Academy of Sciences of France, Companion of the Legion of Honour, and member of almost every Academy in Europe and America which cultivated the Natural Sciences. In 1836 he received the honour of knighthood from His Majesty William IV,

[1] Of which I find no copy in the Club.

together with the insignia of the Order of the Guelphs of Hanover, then an appanage of the British Crown.

In evidence of the estimation in which my father was held by his botanical contemporaries, I think I cannot better conclude this sketch of his life and labours than by giving the following extracts from the obituaries of him drawn up by the two most eminent then living botanists, one in America, the other in Europe. Of these, Prof. Asa Gray thus writes in the 'American Journal of Arts and Sciences,' 2nd Series, xli. 1 (1866):—'Our survey of what Sir William Hooker did for science would be incomplete indeed if it were confined to his published works—numerous and important as they are—and the wise and efficient administration through which, in a space of twenty-four years, a Queen's flower and kitchen garden and pleasure grounds have been transformed into an imperial botanical establishment of unrivalled interest and value. Account should be taken of the spirit in which he worked, of the researches and explorations he promoted, of the aid and encouragement he extended to his fellow labourers, especially to young and rising botanists, and of the means and appliances he gathered for their use no less than for his own.

'The single-mindedness with which he gave himself to his scientific work, and the conscientiousness with which he lived for science while he lived by it, were above all praise. Eminently fitted to shine in society, remarkably good-looking, and of the most pleasing address, frank, cordial, and withal of a very genial disposition, he never dissipated his time and energies in the round of fashionable life, but ever avoided the social prominence and worldly distinctions which some sedulously seek. So that, however it may or ought to be regarded in a country where Court honours and Government rewards have a fictitious importance, we count it a high compliment to his sense and modesty that no such distinctions were ever conferred upon him in recognition of all that he accomplished at Kew.

'Nor was there in him, while standing in a position like that occupied by Banks and Smith in his early days, the least manifestation of a tendency to overshadow the science with his own importance, or of indifference to its general advancement. Far from monopolizing even the choicest botanical materials which large expenditure of time and toil brought into his hands, he delighted in setting other botanists to work on whatever portion they wished to elaborate; not only imparting freely, even to young and untried men of promise, the multitude of specimens he could distribute, and giving to all comers full access to his whole herbarium, but sending portions of it to distant investigators, so long as this could be done without too great detriment or inconvenience. He not only watched for opportunities for attaching botanists to Government expeditions and voyages, and secured the publication of their results, but also largely assisted many private collectors, whose fullest sets are among the treasures of far the richest herbarium ever accumulated in one man's lifetime, if not the amplest anywhere in existence.'

From Prof. Alphonse de Candolle's long *éloge* (La vie et les écrits de Sir W. Hooker), published in the 'Archives des Sciences de la Bibliothèque universelle de Genève,' January, 1866, I have taken the following passages:—

'Et ici je me plais à répéter ce que beaucoup d'autres ont dit ou écrit. Hooker n'était pas de ces hommes qu'on oublie quand on les a vus une fois ou deux. Ses manières étaient aisées, affables, sa complaisance était réelle, son hospitalité charmante. La grâce de Lady Hooker y ajoutait beaucoup, j'en conviens, de sorte qu'il restait de la plus courte visite une impression durable. Sir William m'a toujours paru un type de vrai *gentleman* anglais. Il en avait les bonnes qualités et il en acceptait les charges. Poli envers tout le monde, libéral, oubliant ses intérêts au profit de la science, répondant à toutes les lettres et à toutes les demandes, il avait obtenu dans l'opinion publique une position exceptionnelle. Il était le protecteur des jeunes botanistes et des nombreux amateurs d'histoire naturelle qui partaient pour les colonies. S'il fallait créer un établissement public, donner des subventions, les ministres le consultaient. Sous ce rapport son influence directe ou indirecte s'est fait sentir dans le monde entier. Si l'on publie actuellement des Flores de presque toutes les colonies anglaises, on le doit principalement à ses conseils.

'Maintenant, dans quelle classe devons-nous ranger Sir William Hooker? Evidemment dans celle des botanistes actifs.

'Je pose la plume. Je parcours les rayons de ma bibliothèque, uniquement composée de livres de botanique et assez considérable, je consulte l'ouvrage précieux du *Thesaurus literaturae botanicae* de Pritzel, et je m'adresse la question suivante: en laissant de côté les compilateurs, quels sont les botanistes qui ont le plus écrit? La diversité des formats, la multiplicité des éditions, le mélange dans quelques ouvrages de morceaux de plusieurs auteurs, enfin la dispersion dans les journaux empêchent de répondre à cette question avec toute la précision désirable. Il me semble cependant que Linné, Augustin-Pyramus de Candolle et Sir William Hooker sont les trois botanistes qui ont été le plus laborieux.

'Peu de botanistes ayant eu à nommer des espèces de tous les pays, surtout des espèces de jardins, ont fait aussi rarement que lui des erreurs. Il avait eu soin de s'entourer de riches herbiers et d'une grande bibliothèque ; il avait bonne mémoire; son coup d'œil était rapide. Grâce à tout cela ses descriptions marchaient vite et bien. On est rarement appelé à transporter d'un genre dans un autre les espèces qu'il a classées. Celles qu'il dit être nouvelles, le sont véritablement, à de rares exceptions près, et c'est un degré d'exactitude assez difficile à atteindre en ce qui concerne les plantes cultivées, et dans un ouvrage paraissant à jour fixe, comme le *Botanical Magazine*.

'Je me suis permis de caractériser nettement le botaniste en ce qui concerne ses travaux. Mais il y a, ne l'oublions pas, à côté des ouvrages de Hooker, l'action généreuse, incessante et éclairée qu'il a su exercer autour de lui et à distance. Il a inspiré le goût de la botanique à une foule de personnes, en particulier à son fils. Il a organisé l'établissement scientifique et horticole de Kew, un de ceux où l'on travaille le plus et où les botanistes de tous les pays trouvent le plus de ressources. En arrangeant et en démontrant les belles collections de ce jardin, il a rendu la science populaire. Grâce à ses antécédents et à l'agrément de ses manières, il a obtenu beaucoup en faveur soit de la botanique soit de l'horticulture. Ses recommandations étaient puissantes, même en Australie, dans l'Inde ou en Amérique, de telle sorte que bien des voyages, bien des découvertes et beaucoup de publications importantes sur la flore de pays lointains se rattachent déjà ou se rattacheront à lui, au moins par leur origine.'

J. D. HOOKER.

Nov. 1902.

Printed in the United States
By Bookmasters